Contents

Introduction	2
For a workers and farmers government in the United States *Jack Barnes*	3
Appendix I: Introductory note to 'Workers and Farmers Governments Since the Second World War' *Joseph Hansen*	52
Appendix II: Letter from Joseph Hansen to Robert Chester (1975)	54
Appendix III: An exchange of letters between Joseph Hansen and Robert Chester (1969–70)	57
Appendix IV: On the character of the Algerian government, 1964 statement by United Secretariat	69
Appendix V: From Blanqui to Moncada *Joseph Hansen*	70
Appendix VI: Resolution on workers and peasants government adopted at the June 1923 ECCI Plenum	72
Appendix VII: Zinoviev's report at the June 1923 ECCI plenum	76
Appendix VIII: The social transformations in Eastern Europe, China, and Cuba *Joseph Hansen*	78
Appendix IX: From 'Two Proposals' *George Breitman*	88
Appendix X: A proposed change in transitional slogans, 1967 motion by SWP Political Committee	89

Introduction

The report, "For a Workers and Farmers Government in the United States" by Jack Barnes, was first adopted by the National Committee of the Socialist Workers Party in March 1982, and subsequently by the August 1984 convention of the SWP. Jack Barnes is national secretary of the Socialist Workers Party and a leader of the Fourth International.

The ten appendices in this publication are all referred to in the report by Barnes, which will provide necessary background information about them; also included at the beginning of each one are short explanatory notes. Several of the appended items are by Joseph Hansen, a longtime leader of the SWP and Fourth International who, prior to his death in early 1979, wrote extensively on the question of the workers' and farmers' government. This was part of discussions in the Socialist Workers Party and Fourth International about the post–World War II revolutionary victories in Eastern Europe, China, Cuba, and Algeria. A more extensive selection of Hansen's articles on this topic can be found in another Education for Socialists publication, *The Workers' and Farmers' Government* by Joseph Hansen. This and other related materials can be ordered from www.pathfinderpress.com.

Related articles concerning the questions dealt with in the report by Jack Barnes and the appendices to it have appeared in two issues of the magazine of Marxist politics and theory, *New International*. These articles are: "Forging a Fighting Worker-Farmer Alliance: The Answer to the Crisis of Working Farmers" by Doug Jenness and "The Workers' and Farmers' Government in the United States: An Alliance of the Exploited Producers" by Jack Barnes, both published in *New International* no. 4; and "The Workers' and Farmers' Government: A Popular Revolutionary Dictatorship" by Mary-Alice Waters in issue no. 3. See also *Their Trotsky and Ours: Communist Continuity Today* by Jack Barnes, published by Pathfinder. All three can be ordered from www.pathfinderpress.com.

Steve Clark
JULY 1985

FOR A WORKERS AND FARMERS GOVERNMENT IN THE UNITED STATES

By Jack Barnes

The general line of the following report was adopted by the National Committee of the Socialist Workers Party on March 1, 1982 by a vote of: regular: 38 for, 2 against, 1 abstention, 0 not voting; consultative: 38 for, 0 against, 1 abstention, 0 not voting.

I. THE POLITICAL FRAMEWORK

Working people in the United States today face a social and economic catastrophe greater than at any time in their memory. Washington and its imperialist allies are increasingly resorting to use of their military power as they head toward more and wider wars. As the current bipartisan capitalist government becomes more and more responsible in workers' minds for this unfolding catastrophe, they are increasingly open to explanations of the need for a radical change in the kind of government.

Plants are closing down. Unemployment is at the highest levels since the depression. Farmers face more foreclosures than at any time since the 1930s.

Rising prices continue to slash away at our buying power. Social services and the most elementary social gains continue to be brutally cut by the employers' representatives in the White House and Congress. Giveaway contracts continue to be forced on auto workers and other industrial workers, under the false pretext that this will save jobs and keep factories open.

Just this morning, "laughing boy" Douglas Fraser, president of the United Auto Workers union, signed the renegotiated contract with Ford, giving up pay and job control in exchange for alleged job security. But of course, there's no way that capitalism can guarantee job security. Fraser doesn't even pretend he's sad about these concessions. He just laughs during the televised news conferences, as he sits there beside Ford executives, signing away workers' jobs, income, and working conditions.

Blacks, Latinos, and Asians are confronted by an offensive against school desegregation, busing, language rights, voting rights. Inequality becomes greater, not narrower. Racist violence and police brutality are escalating. Affirmative-action gains are being dismantled. Foreign-born workers are being targeted for stepped-up raids by *la migra* and new anti-immigrant legislation. Democratic rights across the board are under attack.

Women's rights are on the chopping block, too. Abortion rights are being steadily eroded, and the ERA is going down to defeat at the hands of Democratic and Republican politicians.

We should remember that people judge their well-being not by some absolute, ahistorical standard. Working people have carried out hard-fought struggles in this country, and they have developed certain expectations of what their rights are and should be—social, economic, and political rights. They are now seeing those rights being taken away, and their anticipation of further gains being shattered, by the government and by the employers whose interests that government advances.

The increasingly parasitic character of the capitalist economy in its decline, the growing financial speculation, is another barometer of the coming catastrophes. Workers, in general, are much more attuned to this phenomenon than public opinion polls give them credit for. Workers sense that something basic is out of control in the economy when they watch these frantic exchanges of paper taking place; when big corporations that can't or won't retool instead spend billions to buy up stocks and bonds in other companies; when the prices of rare commodities soar and paintings are sold for millions of dollars; when land speculation increases; when personal and business bankruptcies reach their highest levels in decades.

Today we open the newspaper and read that the savings and loan network is on the verge of disaster, with many institutions going under or being forced into mergers. If you read the articles carefully, you also discover that your money in a savings and loans institution is not really "fully insured," as it says on the bank's front window. You'll get your money back only so long as the funds hold out in the Federal Savings and Loan Insurance Corporation.

What's going on in this country today is not "reindustrialization." Factory utilization has dropped to between 65–70 percent, and big business is not investing heavily in new plants and equipment. The productive investment in new technology that is going on eliminates rather than creates jobs, and even investment along these lines—"robotization" and so forth—is severely restricted by the current stagnation.

The biggest owners of the biggest banks seem to suck in more and more values with less and less renewal of productive capacity—like a fungus growing on a living organism. The only people who seem very happy right now are the bankers, who are raking in sky-high interest—including on bonds to the biggest corporations. Workers can't afford to take out loans at the current interest rates, even if they're among the lucky few who "qualify."

Interimperialist competition

The competitors of American capitalism are prime targets of the high interest rates, tight money policies, and other policies being followed by Wall Street and Washington. Behind the pressures by the Reagan administration on the governments of Western Europe and Japan to adopt trade and banking "penalties" against Poland and the Soviet Union is an attempt by the U.S. capitalist rulers to get another edge on their chief competitors. Washington insists that France and West Germany halt ongoing deals over the Soviet natural gas pipeline, for example, while it continues lucrative U.S. grain sales to the USSR.

Washington is also pressing the Western European and Japanese governments to increase their war spending. This, too, has to do with the competitive drive of U.S. imperialism, not only with military and political considerations.

The competitive conflicts between the American capitalists and their rivals today are such that what Wall Street and Washington really need is an interimperialist war. That is what all modern history teaches us—that the U.S. capitalist class today needs an interimperialist war in order to reassert its economic, social, and political supremacy. It needs to use its massive strategic military edge in the capitalist world to take on its more productive competitors and kick them around.

Why, then, doesn't the U.S. ruling class go to war with Japan? The American capitalists could certainly take care of their Toyota problem that way. But *can* Washington go to war against Japan? What class is going to walk in and pick up the pieces?

The major competing imperialist ruling classes can't go to war with each other. This is precluded by the gigantic shift in the world balance of class forces since the Second World War—by the survival of the Soviet workers state, by its nuclear counterforce to that built up by Washington, and by the toppling of capitalism in more than a dozen countries.

Thus, American imperialism in its decline is unable to directly use its one big edge—its strategic military power and size—against its competitors. At the same time, these competitors have gained major ground on U.S. capitalism on the economic plane over the past two decades, and continue to narrow the gap.

Washington also needs a victorious war against the colonial revolution. This should be no mystery either. *The war has begun.* There is a war in Nicaragua and El Salvador today against the extension of the socialist revolution in this hemisphere. The real question is whether or not the U.S. rulers can *win* it. That will be decided in struggle. Everyone knows that more is at stake than Nicaragua and El Salvador, too. There's Guatemala, Honduras, the whole region from Mexico to Colombia. And other wars can break out overnight.

The growing fear among the American people that they are being led into another big war—any one of which could end in a nuclear nightmare—rivets their attention on the government. What kind of government is it that is plunging us into another Vietnam? That wants to spend billions more on weapons of mass annihilation?

Workers need political answers

This political situation helps throw a spotlight for working people on the question of the government, its role, and whose interests it serves. Big decisions are political decisions, and those that aren't have to become political decisions. So, the proposal that some radically different kind of government is needed no longer seems like a wild idea to millions of people in this country.

This is the big picture that we must put at the center of everything we're saying and doing. There are no individual solutions. Working people also increasingly run up against the limits of what can be won through trade-union action alone, even on issues normally thought of as trade-union questions. Workers are *not* going to be able to save jobs, protect wages, or preserve any small element of control on the job solely within the framework of trade-union action. That's absolutely necessary, but it is not enough.

That's why we advocate a labor party based on the unions. Workers must be able to wage their struggles on the political level and put a new kind of government—one that advances *our* interests—in charge of things.

There are no non-working-class solutions to any of the problems we face—including the threat of nuclear annihilation—that are in the interests of working people. That is what we have to help other workers and the unions to understand.

This is the framework in which we talk about the question of the government in our press, election campaigns, in the unions and on the job, in all arenas of our political activity. This is the framework in which our program today presents the kind of government that working people need. It's this perspective that lies behind the kind of struggles we support and help organize; the kind of methods we advocate and use; the kind of political organizations and alliances that we support and urge be formed.

Search for political solutions

It's not just the SWP that is thinking about questions such as these. Others with whom we work face exactly the same political issues, and they are pushed to look for political solutions, for a new kind of government.

This is the case for the National Black Independent Political Party. If you're serious, you can't form a political party without confronting the question of governing. That's what political parties do if they're serious. They try to win in order to govern. One of the aims of a political party may be to better organize struggles and so on, but the question of governing flows naturally from the party's very existence.

But how are you going to govern? What program are you going to implement? In whose interests? What kind of government do you want? With what aim? Is the goal simply to get elected to Congress, as it now exists?

NBIPP has already adopted a charter that opposes imperialism and capitalism, a charter that champions the interests of Black workers and farmers and other working people. NBIPP is going to have to give some thought to the governmental question, too, and arrive at an answer consistent with its charter.

This question came up over and over again during the trial in our lawsuit against the U.S. government last year, as well. The judge kept asking us, "What kind of government do you want to set up? How are you going to do it? Who is it going to represent?" And we took the opportunity to explain.

Socialists present the question of the government concretely—as something flowing from a plan of action that can ultimately solve the crisis facing working people under capitalism.

What kind of government do working people need if we are to be sure we won't wake up some morning and find that U.S. troops are dying in El Salvador? Or that Nicaragua or Grenada have been bombed? Or that four more plants have been shut down? Or that the international monetary system has collapsed? Or that food is not getting into some city somewhere? Or that we've been drinking poison out of the tap because of industrial pollution?

Socialists believe that a different social system is needed to guarantee that these kinds of catastrophes won't occur. But there has to be a link to this new social system—a governmental link. The current government advances the interests of those who benefit from the existing social system responsible for all these ills. It uses the force and violence necessary to defend that private profit system.

What kind of government, then, is needed to organize and lead the big majority to abolish that system and replace it with one based on human needs?

The governmental question is also posed for those individuals and social layers heading away from our class and its allies—liberals, many middle-class radicals and social democrats, and so on. They, too, are impelled by the crisis to think more about the government, but they are drawing conclusions opposite from ours. They find more and more positive things to say about the possibilities of reforming the existing capitalist government in this country.

Susan Sontag, for example, is discovering the virtues of this kind of government relative to those in countries where—as she recently put it—workers and farmers sing "The Internationale." Under the pressures of class polarization, the petty-bourgeois layers typified by Sontag face up to the dreary but necessary conclusion that as bad as this government is (and, oh, they do find it bad and very illiterate and ever so ignorant and boorish), it is nonetheless superior to any other kind.

Socialist election campaigns

Explaining the kind of government socialists advocate plays a central role in our election campaigns. Comrades who are or have been candidates know what a challenge it is to find ways to present our transitional program as the program of a government—the program we'd help lead the workers and farmers to carry out if we were the government majority. It's the program we advocate for a mass labor party based on the unions—a party that would fight for a government that could implement that program.

There is a passage in the *Transitional Program* where Trotsky speaks of "those transitional demands which should, in our opinion, form the program of the workers' and farmers' government" (*The Transitional Program for Socialist Revolution* [New York: Pathfinder, 1973, 1977], p. 175 [2014 printing]). That's one of the good short definitions of the transitional program—the program that we advocate a workers and farmers government carry out.

How can our election campaigns get this idea across and tie it to the perspective of struggle? Just as with our use of the labor party slogan, we want to explain the kind of government we propose not as an electoralist or parliamentary solution, but as part of our perspective of revolutionary mass action by the working class and its allies along the road to winning political power.

Our governmental slogan expresses our internationalist perspective as well. There is no wall separating the domestic and foreign policy that we advocate for such a government. There can't be. Capitalism is a world system, and our program is a perspective for the world socialist revolution.

Trotsky wrote to the French comrades in 1934 that the transitional program is "a system of measures that, with a workers' and peasants' government, can assure the transition from capitalism to socialism" (*Leon Trotsky on France* [New York: Pathfinder, 1979], p. 75 [2012 printing]). That is, a program that takes the toilers along the road to a society that produces no more wars, no more social catastrophes, no more horrors of class society.

That's the challenge that humanity faces now, a challenge that only our class, the working class, can lead it in conquering. That's what's so important about the way we use our governmental slogan and the way we connect it—*today* in *this* country—with a program and strategy to solve the problems for working people created by the capitalist crisis.

II. WHAT IS THE WORKERS AND FARMERS GOVERNMENT?

What is this workers and farmers government that we advocate?

Joe [Hansen] used to point out that most Marxists haven't really been very interested in this crucial question. The last thing Joe wrote on the workers and farmers government, and one of the best things he ever wrote on it, was a one-page introduction to an Education for Socialists bulletin written by Bob Chester entitled *Workers and Farmers Governments Since the Second World War*. [See Appendix I.]

This brief introduction, which Joe wrote in 1978, opened with the statement:

> This study deals with a subject that too many socialist militants might appear at first sight

as hardly of great concern.

[But Joe explained that,] The importance of the question becomes obvious when it is thought through and the consequences for political practice are grasped.

Nonetheless, it is a fact that it remains a field of prime interest only to advanced revolutionary cadres. This holds true for the world Trotskyist movement as a whole.

Joe pointed out that something much more than a correct label or characterization was involved in this question. The problem is not just one of *recognizing* a workers and farmers government when it comes into existence through a revolution, as important as that is. The key task for revolutionists is seeking to understand the dynamics of such a government so that we can relate to it, help to advance it, and learn from it—the better to inspire working people to emulate the example here.

First, then, what is a workers and farmers government? Joe's 1978 introduction presented a concise and unambiguous answer: "the first form of government that can be expected to appear as the result of a successful anticapitalist revolution."

Not just in *some* countries, not just in *backward* countries, not just with *inadequate* leaderships, but "the first form of government that can be expected to appear as the result of a successful anticapitalist revolution." Period.

This is the conclusion that we had reached by 1978, as a result of thinking about and generalizing the lessons from workers and farmers governments established since World War II—lessons that the party had begun to draw in the world report presented for the Political Committee by Joe and adopted at our 1969 convention. [See *The Workers and Farmers Government*, by Joseph Hansen, an Education for Socialists publication, Pathfinder, 1974, pp. 29–39 (2005 printing)].

Resolution of dual power
Second, a workers and farmers government comes out of a mass struggle for power which resolves a situation of dual power in favor of the workers and farmers. The armed might of the old ruling classes is dispersed, and another power based on the toilers replaces it. Only with the resolution of dual power do you have "a successful anticapitalist revolution."

Third, a workers and farmers government can only come into being through a revolutionary upheaval, a real people's revolution. It cannot come about in some "cold" way. It cannot simply be imposed by a bureaucracy, by a military caste, or by an existing government. It cannot be voted in by elections, although elections to councils of the toilers or to a constituent assembly may take place along the road of such a revolutionary struggle, or some time following its triumph.

It takes a revolution to bring a workers and farmers government into existence. It takes mass mobilizations. Guerrilla warfare can play a major role in such a revolutionary upheaval. But the masses of toilers themselves must be mobilized sufficiently to crush enough of the resistance of the old ruling classes and their state machinery to resolve the question of dual power. Only then can the workers and peasants actually begin governing.

If the workers and farmers government is led in a revolutionary manner and meets the challenges posed by the relationship of class forces and material conditions it faces—as is happening in Grenada and Nicaragua today—then it is the key instrument by means of which the toilers can expropriate the exploiters, establish state property, and begin planning the economy—that is, consolidate what we call a workers state.

Independent of the bourgeoisie
Fourth, a workers and farmers government is *independent* of the bourgeoisie, but at the same time still *stands on* capitalist economic relations. That's what it inherits, and that's what it stands on to begin with and, to one degree or another, for some time afterwards.

A revolutionary government can't simply *decree* the disappearance of capital. It can try, but it won't work and will create needless chaos. The key, as Joe used to say, is the tendency and direction—the motion by the workers and farmers government toward leading the masses to destroy capitalism and institute a planned economy. As the workers and farmers government leads this process forward, its own base shifts from the capitalist economic relations it inherited to new economic forms and institutions that come out of the expropriation of the exploiters.

Joe wrote about this in a letter to Bob Chester in 1975, commenting on an initial draft of Bob's *Workers and Farmers Governments Since the Second World War*. This letter summarizes the experiences of Michel Pablo in Algeria. [See Appendix II.]

Joe wrote that a workers and farmers government begins "on the basis of the capitalist economy and even part of the capitalist state structure"— this is what all revolutions inherit to one degree or another and cannot get rid of over night. Smashing the state is not the same as vaporizing it. If there were no bridges, no transitions, no dialectics, then life would be simpler. There would be no algebra to learn, only a little arithmetic. But life is not that simple, especially in the course of revolutions.

Anyway, let's get back to the letter to Bob Chester. Joe indicated

> the role that can be played by a workers and farmers government—beginning on the basis of the capitalist economy and even part of the capitalist state structure, destroying and replacing the capitalist state structure, establishing a monopoly of foreign trade, expropriating the key industries, introducing a planned economy, etc., and finally ending up as a regime (good or bad) [—depending on the capacity of the leadership—] standing on a workers state.

That's the role that a workers and farmers government can play. If the masses are *not* led forward to carry out those steps, of course, then the workers and farmers government will not play that role. It will rot out and be dispersed, instead. A historical opportunity will be missed. The door will be opened for restructuring and reconsolidating the state on the basis of capitalist class relations and state machinery.

Joe dealt with the question of the relationship between a workers and farmers government and its leadership in another letter to Bob Chester—an earlier one, written in 1970. [See Appendix III.] This letter also deals with one way that we can misread the October 1917 revolution in Russia. We can forget that there, too, the revolutionary, Bolshevik-led government operated for some time on the basis of the capitalist economy that fell into its hands, until the exigencies of civil war and imperialist invasion forced it to embark on large scale expropriations more quickly than it had planned.

Bob Chester had written Joe and asked, "Above all, what are the dynamics of a worker and peasant regime that make it the 'link in the revolutionary process'?"

Here was Joe's answer:

> What is involved is governmental power. A party or team that gains governmental power thereby gains the possibility of smashing the old state structure and overturning capitalism.
>
> If a revolutionary-Marxist party exists, and gains governmental power under the impulsion of a revolution, there is no question as to the subsequent dynamics. The party assures it through its program, through the cadres imbued with that program, and through the experience gained in the living class struggle that finally puts it in power.
>
> The course of the Russian revolution is a classic example. Note well, however, that the Bolsheviks held power for a period on the basis of the capitalist state structure and the capitalist economy. Time was required to carry out their program. If anything, they had to carry through these changes *prematurely*. (This had to be paid for later, as Trotsky explained, by the New Economic Policy.) Thus the Russian revolution provided the world with the first example of a "Workers and Peasants Government" in power with the task still before it of actually establishing a workers state.

Petty-bourgeois government

We've often referred to workers and farmers governments as radical petty-bourgeois governments. What do we mean by this?

First, we *don't* mean a government that, in its deeds and orientations, advances the class interests of the petty bourgeoisie. We don't mean a government that seeks to promote the expansion of petty commodity production and trade at the expense of and to the detriment of the expansion of state property.

A government that is petty bourgeois in that

sense would not be characterized by deepgoing anticapitalist measures, while such measures are a *key* determinant of a workers and farmers government. As we put it in the "Draft Theses on the Cuban Revolution" adopted by the SWP National Committee in January 1961, a workers and farmers government must display both the

> tendency to respond to popular pressures for action against the bourgeoisie and their agent [and the] capacity, for whatever immediate reasons and with whatever hesitancy, to undertake measures against bourgeois political power and against bourgeois property relations. The extent of these measures is not decisive in determining the nature of the regime. What is decisive is the capacity and tendency. (*The Workers and Farmers Government,* by Joseph Hansen, p. 5.)

Without that, you don't have a workers and farmers government. *With* that, however, you don't have a government that, in its deeds, is advancing the class interests of the petty bourgeoisie. Instead, you have a government that is advancing the class interests of the workers and other exploited producers. So, a workers and farmers government is not petty bourgeois in this sense.

The 1961 "Draft Theses on the Cuban Revolution" also indicated a sense in which such a government *can* sometimes be considered a petty-bourgeois government. It said that one feature that was being pointed to in Cuba was the "radical pettybourgeois background and composition" of the leadership team in the revolution. This relates to a question that I will come back to later in the report—the theoretical problem that we faced in the postwar period as the result of the unexpected victory of anticapitalist revolutions in a number of countries under leaderships that were neither proletarian in composition nor stated program. Yet these governments—workers and farmers governments—ended up mobilizing the toilers to carry out sweeping revolutionary measures along the historic line of march of the proletariat, culminating in the establishment of workers states.

There is another sense in which such a government can be called petty bourgeois, the only sense that holds true for all of them. That is the fact that the job such a government must accomplish in establishing the domination of proletarian economic forms is not yet done. As long as that job is not completed, there is no way it can base itself on something different from the bourgeois economic forms it inherited, even if increasingly diluted with a "mixture" of the proletarian economic forms it is heading towards, i.e., state property. It is a regime in transition from, and thus necessarily straddling, bourgeois and proletarian economic forms; it is petty bourgeois in this sense. The outcome of that transition is not foreordained; it can falter and fail.

In speaking of proletarian economic forms, we're not talking about the construction of socialism; that transition to a society of associated producers is a more prolonged one that can only be carried out on a world scale. We're talking about a planned economy based on state property. What exists in the Soviet Union, China, Eastern Europe, Mongolia, Korea, Vietnam, Cuba; and what is in the process of being consolidated, for example, in Nicaragua and Grenada.

In this fundamental scientific sense, calling a workers and farmers government a petty-bourgeois regime does not necessarily have anything to do with the class character of its leadership.

That depends on the caliber of the dominant party or groupings involved in the government, the relationship of class forces inside it, the give and take between those forces, and how broad an alliance the government is based on.

Few would challenge that the Bolshevik party, for example, was a proletarian leadership. Moreover, leaderships can and do evolve, sometimes changing their class character substantially in the process, as we saw in the course of the Cuban revolution.

Workers' most powerful instrument

The fifth thing we can point out about the workers and farmers government is that it is the most powerful instrument the working class can wield as it moves along its line of march toward establishing a workers state. It is a more powerful instrument than the strongest union imaginable, or the strongest political party. It involves the use of governmental power by the working class to expropriate the ruling capitalist class and to begin rebuilding society on this new foundation.

Of course, that doesn't exhaust the question. A vanguard party of the working class *must* be built, both to lead the toilers in establishing a workers and farmers government and then in using their governmental power to establish a workers state and to carry out its defense and extension to other countries.

The struggle to establish and advance a workers and farmers government opens up the possibility of constructing such a mass revolutionary proletarian party. History has yet to see a truly mass communist party prior to emergence of such an objective situation.

There are two things we should not lose sight of, in this regard.

First, the toilers in this country will never make a revolution and establish a new government unless, well beforehand, the workers have built a communist party along Leninist lines. We know what kind of party it will take to lead a revolution to victory here and in most other countries. The Leninist strategy of party building is our fundamental guide everywhere.

There's a second thing that should never be lost sight of, however. It is only in the process of the revolution and its subsequent consolidation that the vanguard party is transformed into a party of the great masses of the workers and their toiling allies—as the Bolshevik party was transformed, as the Cuban party has been built and transformed, as the Nicaraguan FSLN and Grenadian New Jewel Movement are now being transformed.

A workers and farmers government, then, creates the optimum conditions for the construction of a massive proletarian party that can lead the transformation to a workers state and the national and international struggle for socialism.

Mass organizations must be built, as well, and they must be armed. The revolutionary army, whose nucleus is forged in the armed struggle carried out in the course of the revolution, must be strengthened, expanded, and professionalized. Militias based in the workplaces and neighborhoods, formed in embryo during the mass popular uprising, must involve growing numbers from every layer of the toilers and their allies.

Whatever the form and origin of the popular organizations—whether they rise beforehand, in the midst of the revolution, or are organized afterward, they must move forward and be consolidated by the workers and farmers government.

Popular programs must be initiated to improve the living conditions of the toilers. Measures dealing with the prices of food, housing, and other basic commodities; unemployment; education; health care; workers control in the factories; combating racism, eliminating all racist practices, and implementing real affirmative action; nationalizing the land and cancelling the debts of the working farmers as part of a thoroughgoing agrarian reform; radically changing the foreign policy of the government to advance the world revolution and the well-being of workers and farmers in other countries. All kinds of immediate, popular measures, which will vary from country to country depending on concrete conditions and level of development.

The actual implementation of the transitional program, as Trotsky says, is the task of a workers and farmers government; and carrying out that program to the end is the process of making the transition to the dictatorship of the proletariat—that is, to a workers state standing on state property, economic planning, and a monopoly of international trade.

So, the workers and farmers government opens up an entire new dynamic and direction, an anticapitalist dynamic and direction. This is not an instantaneous transformation of the economy; that's not feasible.

A whole range of social and economic measures can be put into effect immediately, of course, such as measures to combat discrimination and enforce affirmative action. Some measures against racism *were* "instant revolutions" in Cuba—in the barbershops of Havana, for example. Before January 1, 1959, most barbers there who considered themselves white would not cut the hair of Afro-Cubans. After the revolution, an Afro-Cuban would walk into a barbershop, the barber would say, "I'm sorry, I'm too busy," and the Afro-Cuban would walk out and return with the first militiaman he could find. Everyone's hair got cut.

With a government that bases itself on the oppressed and exploited, affirmative action is not hard to begin carrying out.

The establishment of a workers and farmers

government also lays the basis for a further proletarianization of the leadership, even the leadership of a revolutionary workers party at the head of the government. You sometimes hear people speak of this process as "building links" with the proletariat, but this word "links" can be a dangerous one. Of course, firm links must be built and maintained between the government, the party, and the mass organizations of the working class and other toilers.

The proletarianization of the party and the government, however, is not a process of building "links" with the working class. *The leadership must increasingly become proletarian in composition.* And that, of necessity, involves dealing with the reality of women in the country, of oppressed nationalities, of the concrete history of the country and make-up of its population.

This is one of the things that Fidel dealt with at the Second Congress of the Cuban Communist Party in 1980. He noted that the number of workers in the party had tripled since 1975, and that the new Central Committee elected there included more workers, more women, and more veterans of the Angolan war and other internationalist missions. This "means that our Party has become more proletarian," he said, "and therefore more Marxist-Leninist and more revolutionary."

This challenge and necessity will face the leadership of every party, of every victorious workers and farmers government, if the revolution is to move forward.

A 'Jacobin team'

Joe dealt with the question of the relationship of the class character of the government and its leadership in 1969 when he wrote a criticism of the first draft of a resolution on Algeria adopted by the United Secretariat in preparation for a meeting of the International Executive Committee of the Fourth International; the second draft was subsequently changed and we voted for its general line. [See *The Workers and Farmers Government* by Joseph Hansen, pp. 23–28; 84–91.]

In 1964 the United Secretariat of the Fourth International had adopted a resolution recognizing that a workers and farmers government had come into being in Algeria. [See Appendix IV.] The first draft of the 1969 IEC resolution, however, dropped this designation of the Ben Bella government between 1963 and its overthrow in 1965. Instead, the draft spoke of a "Jacobin team" around Ben Bella that had carried out a series of anticapitalist measures.

Joe thought it was a big step backward to drop the recognition of the existence of a workers and farmers government in Algeria for two years just because the Ben Bella leadership had proven unable to carry the process through to the abolition of capitalism. Moreover, the proposed replacement—"Jacobin team" did nothing to contribute to political clarity. Referring to this ill-conceived designation, Joe wrote:

> But the benefits [of it] escape me. Did this "Jacobin team" function as a government? The answer, of course, is that it did. What was the class nature of this "Jacobin team" government? The answer is "petty bourgeois." Did it rest on a capitalist state structure? The answer is "yes." Did it nevertheless undertake measures which if pursued to their logical conclusion would have ended in the establishment of a workers state in Algeria? The answer is "yes."
>
> Four questions, along with their necessary answers, are sufficient to establish that so far as *content* is concerned, the label "Jacobin team," as applied to the Ben Bella regime, designates precisely the same phenomena as the label "workers and peasants government." Moreover, this is without insisting on what was widely agreed at the time—how this government was influenced by the Cuban example, and what striking parallels existed between the Cuban and Algerian revolutions.

The term "Jacobin team," Joe thought, was not the best historical analogy. If you want to go back to find the roots for the development of a current like the Castro leadership, he thought, the place to look is not 1793, not the bourgeois revolution in France. A more useful and accurate analogy, Joe thought, was to Auguste Blanqui and to his attempts in the mid-nineteenth century to lead an early revolutionary proletarian current that would put its money where its mouth was. Joe explained this in the question-and-answer period of a class on the Cuban revolution given to the New York branch of the SWP in the late sixties. [See Appendix V.] Somebody asked Joe whether the Castro leadership was revolutionary or centrist; Joe responded that he thought it was revolutionary.

Joe pointed out that Marx and Engels placed Blanqui in the camp of the proletarian revolution, whatever their differences with him over strategy and tactics. If you want to find some historical roots to something like the emergence of the Castro leadership, Joe said, this is one good place to look.

The leadership of the Cuban revolution, which had its origins in a radical petty-bourgeois movement, is a revolutionary proletarian current that has grown in its capacities and Marxist understanding over its more than 20 years at the head of the Cuban workers state.

We learned from the Cuban revolution that a formation such as the July 26 Movement can take power. Joe correctly pointed out that this is unlikely unless that organization has some proletarian experiences and connections from the beginning. He pointed to the role of the urban underground of the July 26 Movement, the role of the sugar workers, the role of the successful general strike that helped clinch victory for the revolution in 1959.

A much more rapid proletarianization of the Cuban leadership, however, began with the victory of the revolution, the establishment of the workers and farmers government, the big mobilizations that culminated by the fall of 1960 in the first workers state in the Americas, and the efforts to extend and defend the socialist revolution since then.

The Nicaraguan FSLN today is not petty bourgeois—neither in its composition nor its program and policies. The establishment of the workers and farmers government in Nicaragua has already deepened, and will continue to deepen, the proletarianization of the leadership of that revolution. The same holds true for the New Jewel Movement in Grenada.

The point is not the exact class composition at any particular point in the evolution of such a current, but the *direction* toward proletarianizing the leadership as an essential task of defending and deepening the revolution.

Here, Joe pointed out, it is *deeds* rather than words that are decisive. He explained this in his 1970 letter to Bob Chester.

"What are the characteristics that make [a particular regime] a 'worker and peasant government' rather than radical bourgeois or peasant?" Bob had asked Joe.

Joe responded:

> I would say that the chief characteristic is its direction of movement. This is indicated by its words (declared program) and its actions. The actions are decisive and we should discount the words if they prove not to coincide with the actions of the government.
>
> Some notable examples are available for study in this respect—Nasser's socialist demagogy, for instance, in contrast to his use of state power to foster a new capitalist class; and in the case of the Cubans the opposite contrast, assurances in the first stage about maintaining [capitalist] property relations while their actions were to the contrary.

No foreordained time limits

The next point about a workers and farmers government is that there is no *a priori* time limit to how long such a transitional formation can last short of consummating the transition to its new economic base, to a workers state. If it is successfully led, and if it is able to postpone a full-scale civil war, then—to use a rough historical analogy—the length of its "NEP" can be a relatively long one. That ultimately depends on the relationship of class forces, both domestically and internationally.

Moreover, there is no advantage in the abstract to the transition being quick, as opposed to more prolonged. It's dead wrong to judge the leadership of a workers and farmers government by "how quickly" it is expropriating the bourgeoisie. The only correct starting point for a proletarian leadership is the particular economic and social relations that exist in a country and that must be transformed, the class alliances with other toilers that must be built and consolidated, and the strengthening of the proletariat in the leadership of that alliance as the class struggle deepens in the city and country—in the face of the threats and pressures of international capital. That's what defines the "correct" pace.

Of course, the concrete evolution of the class struggle, inside and outside the country, will continually bear upon and alter the terms of this equation. How these factors will affect the pace of the transformation, however, and how the leadership of the workers and farmers government responds to them, cannot be determined abstractly.

We've already seen Joe's opinion, expressed in his 1970 letter to Bob Chester, that, "If anything, [the Bolsheviks] had to carry through these changes prematurely"—because of the civil war and imperialist intervention. I'll come back to some of the lessons for the world revolution that the Comintern drew from the need in Russia to implement the NEP (New Economic Policy) following the civil war.

Today we can point to two workers and farmers governments—Grenada and Nicaragua—that are in their third, going on fourth, years. They are being led well, in fact, in an exemplary revolutionary manner. We're neither in any rush to speed up the process, nor are we worried about its pace and outcome. We are mostly concerned about learning as much as we can from these revolutions, as well as about defending them against imperialist military attack.

Of course, a workers and farmers government can falter, fall back, and fail to carry out the transition to a workers state. That's what happened in Algeria. But whether or not this happens does not depend on some abstract yardstick concerning speed, but on the caliber and consciousness of the leadership, the steps to organize and mobilize the masses, the success in constructing a mass-based revolutionary armed forces and militia, the solidity of the worker-peasant alliance, and so on. It was the political vacillation and retreat of the Ben Bella regime in the face of resistance to its initial course that paved the way for the 1965 coup there.

Learning from revolutionary experience

We've seen a good number of workers and farmers governments in this century.

By studying them, especially those with revolutionary leaderships, we can get a concrete feel for the kind of struggles, the kind of measures and actions, and the kind of leadership necessary to establish a workers state. In this sense, Nicaragua and Grenada, like Cuba and the early Soviet Union, are models for the coming American revolution. They are living models for revolutionaries everywhere.

They are models of the types of revolutionary struggles that can lead to the establishment of a workers and farmers government. They are models of a revolutionary leadership of such a government that applies, in practice, a transitional program that deepens the class struggle and leads the toilers forward to the consolidation of a socialized economy.

Of course, the kind of proletarian party, the type of program and caliber of leadership necessary to lead the American workers against the powerful ruling class of the United States are very concrete questions. They are determined by the specific economic and political factors confronting the working class, our allies, and our class enemy in this country.

It's going to be harder to establish a workers and farmers government here. The American workers and farmers are up against the most powerful capitalist class and government in the world. We know this.

But we should be crystal clear that the *class dynamics* involved in the workers and farmers governments in Russia, Cuba, Nicaragua, and Grenada are models that we must study and learn from. These are revolutions that succeeded and progressed *because of,* not in spite of, their leaderships.

In fact, any Marxist today who doesn't try to spend as much time as possible, within the bounds of responsible revolutionary activity, to study and learn all they can *concretely* from Grenada and Nicaragua is not a serious proletarian revolutionist. Here are two places where workers and farmers governments have been established and whose leaderships are trying, *right now,* to lead the toilers forward to the establishment of workers states. This should be of paramount interest to us, since this is exactly what we want to do here in the United States.

Here, too, a workers and farmers government is "the first form of government that can be expected to appear as the result of a successful anticapitalist revolution." That's the kind of government that a labor party in this country must fight for—a government that would carry out a genuine transitional program. As serious proletarian revolutionists, we have to present ourselves as a movement that believes that the working class and its allies can and must *govern.*

That's why revolutionary-minded workers, the kind attracted to the SWP, are inspired by and determined to learn from the examples of Grenada, Nicaragua, and Cuba. They know that they are learning a great deal about the future struggles and tasks of their class right here at home.

Why so little interest?
If studying workers and farmers governments is so important, then why has there been so little interest in this question, even in the revolutionary movement? Why do so many revolutionaries make this error in judgment? Joe also tried to answer this question in his 1978 introduction to Bob Chester's bulletin on *Workers and Farmers Governments Since the Second World War.*

This is what Joe had to say:

> The main reason for this discounting of the question is to be found, I think, in the paramountcy of problems facing small revolutionary organizations in disseminating a revolutionary-socialist outlook among the masses. A better understanding of what is involved can be gained if we single out three general aspects, or phases, of this consciousness-raising process—not forgetting, of course, that in the final analysis they mesh together.
>
> [First, Joe cited] the educational work of bringing the masses to understand that the great social and economic evils they suffer from are consequences of capitalism in its death agony, and that the dilemma facing humanity on a world scale with ever-increasing acuteness is *socialism or barbarism.* The task is preeminent in countries where the program of revolutionary socialism is represented by only small minority movements [—in other words, by organizations such as ours.]
>
> [The second thing that Joe listed is,] the organizational work of building a revolutionary-socialist mass party as the means for meeting the central dilemma. The problem facing small revolutionary groups of linking up with the masses comes under this heading. The task demands doggedness, the utmost attention, and an expenditure of time and effort bordering on fanaticism.

Joe used the word "fanaticism" intentionally. He took advantage of the occasion of this introduction to make clear that one characteristic of worker-Bolsheviks is a doggedness and commitment of time and effort to the revolution bordering on fanaticism. In that sense, we're fanatics. The Cubans are fanatics.

The third aspect or phase of the work of a revolutionary leadership, Joe said, is, "The final push of playing a leading role in the working-class struggle for power when the conditions for this have matured."

Each word was carefully chosen in this sentence. Note that Joe spoke of the "final push of playing *a* leading role in the working-class struggle for power." A party that can lead the workers to power must display the capacity to reach out to other leaderships, even if only for brief periods of time, in the process of establishing the proletarian party's decisive weight in the leadership of the workers and farmers government.

> For periods longer than expected, revolutionists have had to concentrate on the two preliminary phases. The associated tasks are just as difficult as those of the third phase—perhaps more so. The preliminary problems, standing in some instances for years, if not decades, at the top of the revolutionary agenda, can certainly appear to be more real than the question of what form of government might appear as the consequence of a revolutionary victory.
>
> However, in today's highly unstable world, seemingly remote theoretical questions have a way of suddenly imposing themselves in the political arena and demanding answers that can decisively determine the fate of groups and currents bidding for leadership of the working class.
>
> Thus, problems related to the struggle for power cannot be placed in deep-freeze to be brought out "when the time comes." They are with us now, both in the sense of internation-

ally important events on which stands must be taken (the Cuban victory, for instance), and in the sense of gaining a more concrete appreciation of the possibilities in coming struggles.

> Moreover, the struggle for power, along with the accompanying problems and tasks, must be kept constantly in mind. As the goal, that culminating phase dominates our decisions in selecting the means required for its realization.

This observation applies to revolutionists in the United States today, even though the third phase cited by Joe is a ways down the road. Even at the current stage of building the propaganda nucleus of a proletarian party, we in the SWP should not lose sight of where we're going. It "must be kept constantly in mind," since "the goal, that culminating phase, dominates our decisions in selecting the means required for its realization" in *this* country, *today*.

Raising the perspective of a workers and farmers government in our propaganda is not something we can postpone for a far-off distant future; it's something we want to raise with workers and farmers we talk to right now. And in making this question real and relevant, it helps a great deal to be able to point to this kind of government actually functioning in a few places—to Nicaragua and Grenada, and to the gains after two decades of the Cuban revolution.

"That's what socialists would do," we can tell our co-workers and other working people. "That's the kind of government we need, a government of workers and farmers that acts in *our* interests, not in the interests of the bosses, warmakers, and oppressors." And then we lay out our program—the transitional program that a workers and farmers government would implement, and that we fight for now along the road toward a revolution to establish that government.

III. WHERE THE SLOGAN COMES FROM

Where did this understanding of a workers and farmers government come from? Where did we learn it? How have we changed it in light of new experiences and lessons in the national and international class struggle?

What I've outlined in the first part of this report is not identical to how we used and understood the concept at the beginning of the Cuban revolution. Our opinions on the workers and farmers government in 1969, 1975, and 1978 were not identical to our opinions in 1959 and 1960. When you go back and review the materials, you'll see both the continuity and the evolution.

The workers and farmers government is an integral part of our revolutionary continuity. We originally got it—like so much else—from the experiences of the Bolsheviks. The transitional strategy the Bolsheviks used to build their party was grounded in recognizing the centrality of the worker-peasant alliance as the key to the Russian revolution.

Following the February 1917 revolution, as we know, the Bolsheviks raised the slogan, "All power to the soviets!" And they raised this at a time when the big majority in the soviets was still held by the Mensheviks and by the peasant-based Socialist Revolutionary (SR) party.

But the class-collaborationist misleaders of these parties refused to break with their capitalist partners and to begin implementing the demands of the Russian workers and peasants for peace, sweeping agrarian reform, and other radical measures. Following the massive armed workers' demonstration in July 1917, these betrayers joined with the bourgeoisie in a fierce witchhunt against the Bolsheviks and other revolutionary-minded workers and peasants. The Bolsheviks temporarily dropped the slogan of "All power to the soviets," judging that the toilers would have to find some other forms of organization to push aside the obstacle of the Mensheviks and SRs and move forward their struggle for power.

By September, however, the Bolsheviks again saw an opportunity to call for a government responsible to the soviets. Lenin drafted an article suggesting that the Bolsheviks offer what he called a "compromise" to the Mensheviks and the SRs.

> The compromise on our part is our return to the pre-July demand of all power to the Soviets and a government of S.R.s and Mensheviks responsible to the Soviets.
>
> Now, and only now, perhaps *during only a few days* or a week or two, such a govern-

ment could be set up and be consolidated in a perfectly peaceful way. In all probability it could secure the peaceful *advance* of the whole Russian revolution, and provide exceptionally good chances for great strides in the world movement towards peace and the victory of socialism.

In my opinion, the Bolsheviks, who are partisans of world revolution and revolutionary methods, may and should consent to this compromise only for the sake of the revolution's peaceful development—an opportunity that is *extremely* rare in history and *extremely* valuable, an opportunity that only occurs once in a while.

The compromise would amount to the following: the Bolsheviks, without making any claim to participate in the government (which is impossible for the internationalists unless a dictatorship of the proletariat and poor peasants has been realized), would refrain from demanding the immediate transfer of power to the proletariat and the poor peasants and from employing revolutionary methods of fighting for this demand. A condition that is self-evident and not new to the S.R.s and Mensheviks would be complete freedom of propaganda and the convocation of the Constituent Assembly without further delays or even at an earlier date.

The Mensheviks and S.R.s, being the government bloc, would then agree (assuming that the compromise had been reached) to form a government wholly and exclusively responsible to the Soviets, the latter taking over all power locally as well ("On Compromises," *Collected Works*, Vol. 25 [Moscow: Progress Publishers, 1977], p. 310–11).

Under those conditions, Lenin said, the Bolsheviks would peacefully campaign to win a majority for its program in the soviets and seek by these means to establish a revolutionary government that would pursue the task of mobilizing the workers and peasants for the transition to a workers state.

As we know, the Mensheviks and SRs did not break from their policies of subordination to the capitalists, and the Bolsheviks led the workers of Petrograd in a victorious insurrection on October 25.

Even then, as Trotsky explains in his *History of the Russian Revolution*, the Bolsheviks did not reject the idea of a broader coalition government of parties based on the workers and peasants. In the meeting of the All-Russian Soviet of Workers and Soldiers deputies on October 25, the day of the insurrection, the Bolsheviks expressed their willingness to form a government that would include not only the Bolsheviks—by now a majority in the soviets—but also the Mensheviks, Socialist Revolutionaries, and the Left Socialist Revolutionaries.

The Mensheviks and SRs, however, with the exception of the Left SRs, rejected this proposal and walked out of the soviets. They denied the legitimacy of the soviets and the insurrection, and sought to reopen a dialogue with the capitalist Provisional Government. Only after this did leading Bolsheviks altogether rule out a coalition with these betrayers.

As Trotsky explained in a speech before the soviet congress on October 26, after the Bolsheviks had gone "to [the Mensheviks and SRs] and [said], 'Let us take the power together!' they [ran] to the city duma and unite[d] there with open counterrevolutionists! They are traitors to the revolution with whom we will never unite" (*History of the Russian Revolution* [New York: Pathfinder, 1980], Vol. 3, p. 1392 [2014 printing]).

The soviet congress voted to set up a Council of Peoples' Commissars to serve as the government. The Bolsheviks tried to get the Left SRs to participate in this council, and finally succeeded in early December; the resulting governmental coalition had a majority of Bolshevik members and a substantial minority of Left SRs. This was important for the Bolsheviks, since the Left SRs remained the majority party among the poor and middle peasants at the time; at the December 1917 congress of the Soviets of Peasants Deputies, for example, the Left SRs had 350 delegates to the Bolsheviks' 91.

The name adopted on October 26 for the Council of Peoples' Commissars, on Lenin's proposal, was the Provisional Workers and Peasants Government; in January the "provisional" was dropped by the Third All-Russian Congress of Soviets, following the dispersal of the Constituent Assembly.

The Left SRs resigned from the government in March 1918 because of their disagreements over the Brest-Litovsk peace treaty and under the mounting pressure of the class struggle in the countryside between the kulaks and the poorer peasants. In July prominent Left SR leaders participated in planning the assassination of the German ambassador and in an attempted coup against the Bolshevik-led government.

While the majority of the Left SR leaders went over to the counterrevolution, two groups split from the Left SRs—the Narodnik Communists and Revolutionary Communists. They continued to support the revolutionary government, maintained their own newspapers and organizations for a period, and eventually fused with the Bolshevik Party (in November 1918 and October 1920, respectively).

The defection of the Left SR majority occurred simultaneously with the launching of the Committees of Poor Peasants at the Bolsheviks' initiative in the summer of 1918. As the class struggle intensified and differentiations deepened in the countryside, the Bolsheviks won a growing base among the rural proletariat, poor peasants, and layers of the middle peasants, as well.

Fourth Comintern Congress

As Joe points out in his introduction to Bob Chester's bulletin, the first discussion of the workers and farmers government *per se* in the Marxist movement occurred in the Communist International.

Joe remarks that the topic

> came under consideration quite late in the development of key revolutionary-socialist concepts. It was submitted for general discussion for the first time at the Fourth Congress of the Communist International in [November–December] 1922. Only the delegates of the Bolshevik Party, in the period when it was led by Lenin and Trotsky, could have suggested the importance of the question to the cadres of the Third International.

Joe explained that this discussion was based on the Bolsheviks' generalizations from their own and from other revolutionary experiences since 1917:

> The delegates at the Fourth Congress did not engage in fanciful speculation. Their debate was based on the experience of the October 1917 revolution in Russia, on five years of thinking over that mighty chapter in the development of civilization, and on the need to bring subsequent experiences into the context of the lessons of 1917.

The "Theses on Tactics" adopted at the Fourth Comintern Congress contained a section on the use of the workers government or workers and farmers government as a transitional slogan. It outlined the political course that such a government would follow:

> The most elementary program of a workers government must consist in arming the proletariat, disarming the counterrevolutionary bourgeois organizations, installing supervision over production, insuring that the main burden of taxation falls on the rich, and smashing the resistance of the bourgeois counterrevolution.
>
> A government of this sort is only possible if it emerges from the struggle of the masses themselves. [The continued existence of such a government,] carrying out a revolutionary policy must lead to a fierce struggle and, eventually, to a civil war with the bourgeoisie (portions reprinted in *The Workers and Farmers Government,* an Education for Socialists Bulletin, pp. 53, 54; full text available in *Theses, Resolutions and Manifestos of the First Four Congresses of the Third International,* ed. Alan Adler, [London: Ink Links, 1980], p. 397–8).

The Fourth Congress resolution also discussed the approach that communists should take toward participation in such a government. "In certain circumstances," it said, "Communists should declare that they are prepared to form a government with workers parties and organizations that are non-Communist. But they can take such an action only when guarantees are given that these workers governments will really carry out a struggle against the bourgeoisie in the sense indicated above." If that is done, the resolution said, then the workers government or the workers and farmers government can "serve as a point of departure for attaining" the dictatorship of the proletariat.

Trotsky's view

Trotsky played a leading political role for the Bolsheviks at the Fourth Comintern Congress. Subsequently, he spoke at a December 1922 Bolshevik gathering to report back on the results of the congress. In that speech, Trotsky put the discussion and decision on the workers government at the Fourth Congress in the framework of the Bolsheviks' own experience following the October revolution.

> Under certain conditions the slogan of a workers' government can become a reality in Europe. That is to say, a moment may arrive when the Communists together with the left elements of the Social Democracy will set up a workers' government in a way similar to ours in Russia when we created a workers' and peasants' government together with the Left Social-Revolutionaries.
>
> Such a phase would constitute a transition to the proletarian dictatorship, the full and completed one (*First Five Years of the Communist International* [New York: Pathfinder Press, 1953, 1972], Vol. 2, p. 438 [2015 printing]).

Trotsky returned to this question in a June 1923 article in *Pravda,* linking it to the perspective of the European-wide proletarian revolution.

> In connection with the slogan of "A Workers' and Peasants' Government", the time is appropriate, in my opinion, for issuing the slogan of "The United States of Europe." Only by coupling these two slogans shall we get a definite systematic and progressive response to the most burning problems of European development.
>
> To the toiling masses of Europe it is becoming ever clearer that the bourgeoisie is incapable of solving the basic problems of restoring Europe's economic life. The slogan: "A Workers' and Peasants' Government" is designed to meet the growing attempts of the workers to find a way out by their own efforts. It has now become necessary to point out this avenue of salvation more concretely, namely, to assert that only in the closest economic cooperation of the peoples of Europe lies the avenue of salvation for our continent from economic decay and from enslavement to mighty American capitalism.
>
> [For this reason Trotsky explained,] the slogan of "The United States of Europe" has its place on the same historical plane with the slogan "A Workers' and Peasants' Government"; *it is a transitional slogan, indicating a way out, a prospect of salvation, and furnishing at the same time a revolutionary impulse for the toiling masses* (emphasis added).

This transitional sense remains the purpose of our use of a governmental slogan today—to indicate "a way out, a prospect of salvation, and furnishing at the same time a revolutionary impulse" for American workers and farmers.

Trotsky's 1923 article also took up another question relevant to our discussion here today.

> Is the realization of a "Workers' Government" possible without the dictatorship of the proletariat? Only a conditional reply can be given to this question. In any case, we regard the "Workers' Government" as a *stage* toward the dictatorship of the proletariat. Therein lies the great value of this slogan for us (Trotsky's emphasis, *First Five Years of the Communist International,* Vol. 2, pp. 460, 461–62, 463, 466).

Here again, the stress is on the transitional character of the workers and farmers government, both as a propaganda slogan and as a transitional regime leading toward the expropriation of the exploiters.

June 1923 ECCI meeting

At the Fourth Congress, the Comintern had looked at the governmental question primarily from this standpoint as a transitional slogan in connection with the Bolsheviks' united front approach. The terms "workers government" and "workers and farmers government" were thus used somewhat interchangeably.

A June 1923 meeting of the Executive Committee of the Communist International (ECCI) took the discussion further, pointing to the importance of the alliance between the working class and the peasantry as a key aspect of this transitional approach. The resolution adopted at the June 1923 meeting of the ECCI put the question in the context of the discussions at the Second and Fourth Comintern congresses on the worker-peasant alliance. [See Appendix VI.] The resolution pointed out:

> The Communist Parties must not regard themselves as the parties of the extreme proletarian opposition *within* bourgeois society, as was the case during the period of the development of the Second International. The Communist Parties must develop in themselves the psychology of parties which sooner or later will lead the toiling masses into the fight against bourgeois society, to overthrow the bourgeoisie and to replace it as the rulers of the State. The narrow craft psychology must be replaced by the psychology of parties which possess the will to power, which embody the interest of class hegemony in the revolution.
>
> A Communist Party must prepare itself to defeat the bourgeoisie tomorrow and *therefore* today adopt aims common to all the people. It must therefore attempt to attract to the support of the proletariat all those sections which because of their social position, will be able at the critical moment to support the proletarian revolution in one way or another.
>
> The motto of the "Workers' and Peasants' Government," like that of the Workers' Government in its time, does not in any way replace or put in the background the agitation for the dictatorship of the proletariat—the foundation of foundations of Communist tactics. On the contrary, the motto of the Workers' and Peasants' Government, by extending the basis of the tactic of the united front—the only correct tactic for the present epoch—is the path to the dictatorship of the proletariat.
>
> The correct interpretation of the motto of the Workers' and Peasants' Government will permit the Communists not only to mobilise the proletarian masses of the towns, but also to create valuable points of support in the countryside and thus prepare the ground for the seizure of power.

The resolution took note of the need to fashion the application of this transitional slogan to the particular circumstances in which it is used:

> It will, of course, be understood that the agitation, carried on under the slogan Workers' and Peasants' Government must be adapted to the conditions prevailing in each country, for instance in the United States it will apply to working farmers.

The ECCI resolution also explained the relationship between the slogan of the workers and peasants government and the fight for independent working-class political action:

> It is self-understood that penetration into the heart of the masses of the peasantry and the motto of "Workers' and Peasants' Government" by no means imply the conversion of our Party from the workers' party into a "Party of Labour" or a "Workers' and Peasants' Party." Our Party, as far as its composition and aims are concerned, must remain a party of the working class, but of a working class which draws in its wake all sections of the working population and leads them into the fight against capitalism.

Lessons of NEP

Zinoviev's report on this resolution at the June 1923 ECCI plenum dealt, among other things, with lessons the Bolsheviks had drawn from their experience with the NEP since its introduction at the close of the civil war in 1921. He related these lessons to the question of the workers and peasants government. [See Appendix VII.]

> At the Fourth Congress, we explained to you, why in our opinion the New Economic Policy of the Soviet government is an international phenomenon and not merely an episode in the Russian Revolution. We have proved to you that almost every country after the revolution will have to go through a more or less long phase of this type of policy . . . and that the victorious proletariat of any country will have to face the problem of the appropriate unification of the working class

and the peasantry when the time comes.

If this is so, and we do not doubt it, the logical conclusion to be drawn is the necessity of a worker and peasant government.

If we consider conditions in a number of countries, we do not see a single country for which this solution would not be the most fitting. We now say to the backward workers and peasants: we want to destroy the state of the rich, and we want to create a state of workers. Let us decide to add: for this reason we suggest the formation of a worker and peasant government.

The lessons of the NEP, both for the Soviet Union and the Comintern, had indeed been a central focus of the Fourth Congress discussions. Lenin's only speech at that Congress (he was ailing at the time) centered on the NEP. Lenin told the Congress that he hoped

> to draw very important practical conclusions for the Communist International [—conclusions relevant] not only from the viewpoint of a country whose economic system was, and is to this day, very backward, but also from the viewpoint of the Communist International and the advanced West-European countries.
>
> [The introduction of the NEP, Lenin explained, marked the Bolsheviks' recognition that] the direct transition to purely socialist forms, to purely socialist distribution, was beyond our available strength, and that if we were unable to effect a retreat so as to confine ourselves to easier tasks, we would face disaster. [This question, he added, is] of first-rate importance to all the Communist parties (*Collected Works*, Vol. 33, p. 420–22).

Trotsky also presented a major report on the NEP to the Fourth Congress. (See *First Five Years of the Communist International* [New York: Pathfinder Press, 1977].) Lenin subsequently suggested in a letter to some American Communists that "those to whom the question of our New Economic Policy—the only correct policy—is not quite clear" should refer to both his own and Trotsky's reports at the Fourth Congress.

Trotsky's report accompanied a set of "Theses on the Economic Situation of Soviet Russia from the Standpoint of the Socialist Revolution," which Trotsky had drafted. This document, too, pointed to some important general lessons about the transition from property forms inherited from the old ruling class to new proletarian socialist forms based on state property.

> . . . the Soviet government is following an economic path which it would doubtless have pursued in 1918–1919 had not the implacable demands of the Civil War obliged it to expropriate the bourgeoisie at one blow, to destroy the bourgeois economic apparatus and to replace the latter hastily by the apparatus of War Communism.
>
> The most important political and economic result of the NEP is that we have obtained a serious and stable understanding with the peasantry who are stimulated to expand and intensify their work by gaining access to the free market (p. 367).

These lessons from the Fourth Comintern Congress were incorporated into the perspectives of the June 1923 ECCI resolution on the workers and peasants government.

> The slogan of a "Workers' and Peasants' Government" will render good service to the communist parties even after the seizure of power by the proletariat; for it will remind the proletariat of the necessity to harmonise its movements with the sentiments of the peasantry in their respective countries, to establish a correct coordination between the victorious proletariat and the peasantry, and to observe a rational policy in the gradual introduction of the economic measures of the proletariat, such as was arrived at by the victorious proletariat of Russia in that period of the Russian revolution which is called the new economic policy.

While no one in the Comintern could have foreseen the concrete development of workers and farmers governments following World War II, their discussions certainly help prepare us to understand the transitional processes now under way in Nicaragua and Grenada. As I pointed out earlier in the report, there can be no foreordained schedule of how quickly a workers and farmers government will complete the transformation to its new economic base, that is, to a workers state. As the ECCI resolution explained, that depends on concrete economic and political conditions and the particular relationships that exist between the workers, farmers, and other exploited producers. Today, with the experience from revolutions since that time, the truth of this lesson is even clearer.

The Transitional Program

With the bureaucratic degeneration of the Comintern, the slogan of the workers and peasants government—like so much else—began to be misused by the Stalinists as a cover for their class-collaborationist strategy of subordination to the bourgeoisie. Instead of applying the slogan as part of a strategy for anticapitalist revolutions, as the Comintern had projected it in 1922–23, the Stalinist misleaders *counterposed* it—in both theory and practice—to the socialist revolution. This is the content they gave to the slogan during the Chinese revolution in the 1920s and the Spanish revolution in the 1930s, paving the way for crushing defeats of the workers and farmers in both cases.

The thread begun at the Second, Third, and Fourth Comintern congresses and at the 1923 ECCI meeting, linking the worker-peasant alliance, the united front, and the transitional use of the governmental slogan, was picked up again by Trotsky, especially in the mid–1930s in his discussions with French comrades and further in discussions with American comrades leading up to the adoption of the Transitional Program in 1938.

In Trotsky's discussions with leaders of our movement in this country in 1938, he argued with them at some length about adopting the workers and farmers government slogan, as well as re-adopting the labor party slogan that he had argued against earlier in the thirties. Trotsky presented both slogans as key parts of an overall perspective of strengthening and proletarianizing our party as it prepared for the coming war. Given the kind of domestic and international capitalist crisis and catastrophe that existed in the 1930s, and in connection with the fight for a labor party, a revolutionary workers party had to be able to explain to workers and farmers what kind of government could offer a way forward.

In the Transitional Program, Trotsky explained revolutionists raise the workers and farmers government slogan in order to advance class consciousness. The slogan is "acceptable to us only in the sense that it had in 1917 with the Bolsheviks, i.e., as an antibourgeois and anticapitalist slogan. . . ." Trotsky considered the workers and farmers government slogan, used correctly, to be an anticapitalist slogan.

Trotsky continued that revolutionists "in no case [use the slogan] in that 'democratic' sense which the epigones [Stalinists] later gave it." That is, as a slogan to cover up an alliance with those who would betray the independent class interests of the workers and peasants and turn the revolution over to the bourgeoisie.

To use the slogan the first way, Trotsky said, was "a bridge to socialist revolution," while to use the slogan the second way was to transform it "into the chief barrier upon [the] path" of the socialist revolution.

Trotsky's discussions with the American comrades on the slogan are useful as we try to figure out how to use it in our press and election campaigns today. "Large masses will understand it in a democratic parliamentary sense," Trotsky said, "but we will try to explain it in a revolutionary sense." If we reject the workers and farmers government slogan, he pointed out, "then we have deprived ourselves of the possibility of saying to the poor farmers, 'It will also be your government'" (*The Transitional Program for Socialist Revolution;* pp. 127, 174–75, 252).

IV. WHAT IS THE DICTATORSHIP OF THE PROLETARIAT?

We should stop for a moment to make an important point that might initially seem to be a terminological digression. This has to do with the way we use the term "dictatorship of the proletariat" today in our movement, and what its relationship is to the workers and farmers government.

In the Transitional Program, Trotsky expressed the judgment that even if a workers and farmers government should come into existence under a nonproletarian leadership, an eventuality he considered unlikely, it would "represent merely a short episode on the road to the actual dictatorship of the proletariat." In any case, Trotsky added, the sense in which *we advocate* the workers and farmers government is as a popular formulation or synonym for the dictatorship of the proletariat, one that underscores the alliance of the workers and peasants upon which its power is based.

What did Trotsky mean by this? Did he mean that a workers and farmers government was only possible if bourgeois property was expropriated virtually instantaneously after the victory of the revolution? Clearly not. "The fact is that the dictatorship of the proletariat does not at all coincide mechanically with the inception of the socialist revolution," he wrote in 1931.

The problem here is that the term "dictatorship of the proletariat" has been used, and is still often used, *in two different ways* by Marxists.

The first way is synonymous with the new revolutionary power—whether you call it a government, regime, or state—that is brought into being by a successful anticapitalist revolution that resolves the situation of dual power by smashing the old bourgeois state apparatus and army and replacing them with new ones based on the workers and peasants. That new revolutionary power then faces the task of organizing and mobilizing the working class to lead the farmers and other toiling allies and together with them carries out the expropriation of the bourgeoisie and the establishment of what we call a workers state.

This is the way that Marx, Engels, Lenin, and other revolutionary Marxists normally used the term; it is the way Trotsky used it prior to the bureaucratic degeneration of the Soviet Union. Engels referred to the Paris Commune, for example, as the first dictatorship of the proletariat.

In *State and Revolution* Lenin's emphasis was on: 1) the need to carry out a revolution to smash the old capitalist army and state apparatus (a task denied in words by class-collaborationists in the Second International and in deeds by centrists); and 2) the need to construct a new revolutionary state based on the workers and other exploited toilers (a task denied by the anarchists and syndicalists).

Whether or not to take political power out of the hands of the bourgeoisie, and how to construct a new proletarian state to carry through the anticapitalist revolution—these were the central questions that Lenin addressed in *State and Revolution,* basing himself on the writings of Marx and Engels.

How precisely to characterize the transitional government that took power and led the workers in expropriating the bourgeoisie was not a burning question under those conditions. As we've seen, all Marxists at the time believed that only proletarian revolutionists either could or would actually carry out a thoroughgoing revolution to dismantle and replace the old capitalist state. Once that task was completed, it followed that the new revolutionary-led regime would organize the workers to carry out the further tasks—starting with those left uncompleted from the bourgeois revolution, through the socialist tasks of expropriating the bourgeoisie, aiding the extension of the world proletarian revolution, and the eventual transition to socialism on an international scale.

Bolsheviks such as Lenin and Trotsky believed that the failure to spread the socialist revolution would eventually lead to the defeat of the Soviet workers state through a capitalist counterrevolution. Neither they nor other Marxists anticipated that such isolation and imperialist pressure would result in the political expropriation of the workers by a privileged petty-bourgeois bureaucratic caste, but that the economic conquests of the workers state itself—that is, state property, planning, and the monopoly of foreign trade— would survive and determine the class character of that horribly degenerated state.

When you read the works of Lenin from late 1917 through his death in 1924, you will find many different descriptions of the government that was established after the October revolution: dictatorship of the proletariat, dictatorship of the proletariat and poor peasants, soviet government, workers and peasants government, workers and peasants republic, peoples' republic, socialist republic, and others. Since the revolutionary deed had been done in October, and since the socialist goals of the Bolshevik-led regime were clear, arriving at a precise term for the new government was not at issue.

Lenin used whatever term best expressed the aspect of the transitional reality he needed to spotlight, given the particular audience, political opponent, or question he was addressing. Against the centrist Kautsky, who insisted that a proletarian-led revolution had been premature and a revolutionary dictatorship unjustified in Russia, Lenin stressed the character of the regime as a proletarian dictatorship; in other situations, he stressed its character as a republic, as worker and peasant power, or so on.

Two uses

Since the 1930s, however, our movement has usually used the term "dictatorship of the proletariat" in another way. We use it as a synonym for a "workers state"—that is, a state where the dominance of bourgeois economic relations has *already* been abolished and replaced by state property, economic planning, and a monopoly of foreign trade.

Thus, the *same* term, "dictatorship of the proletariat," is used to define two closely related but different things: first, the regime that organizes the workers and their allies to carry out the expropriation of the bourgeoisie; and second, the regime that comes into existence after that expropriation of the bourgeoisie has already been carried out. This obviously can lead to problems. Sometimes, in fact, we try to cram both ideas into the same definition. For example, turn to the "notes" section of many Pathfinder books. There you will find a standard, almost unvarying, definition:

> Dictatorship of the proletariat is the Marxist term for the form of rule by the working class that will follow rule by the capitalist class ("dictatorship of the bourgeoisie"). More modern substitutes for "dictatorship of the proletariat" are "workers state" and "workers democracy".

Although this is fine as a rough-and-ready definition for most occasions, it leaves out something crucial: the *transition* from the capitalist state to the workers state. In doing so, this definition glosses over the distinction between government and state, and the contradiction that is inevitable for a longer or shorter period of time (depending on the concrete circumstances) between the class character (1) of the new government and (2) of the economic property forms the state rests on—forms the new government inherits and on which it is based during the transitional period following a successful seizure of political power by the proletariat. Educating, organizing, and mobilizing the workers and peasants to achieve a resolution of that contradiction in a historically progressive direction is the central task of a workers and farmers government.

The Cuban, Nicaraguan, and Grenadian revolutionists usually call this moving to socialism; we call it establishing a workers state. We're all talking about the same transformation in a given country, the successful expropriation of the bourgeoisie and the development of governmental forms in harmony with this new economic base.

Degeneration of Soviet workers state

This seemingly inconsequential question of exactly what communists mean by the "dictatorship of the proletariat" became a pressing political question in the 1930s with the degeneration of the world's first workers state, the Soviet Union. In the fight with the petty bourgeois opposition in our party at the end of the 1930s, it marked the dividing line between defense of the Soviet workers state and capitulation to world imperialism—between revolution and counterrevolution.

The Soviet working class had been politically expropriated by a privileged, petty-bourgeois caste; the *government* under Stalin and his heirs became a dictatorship of the bureaucracy, not of the working class and peasantry, in that sense. *But the dictatorship of the proletariat, the workers state, survived, since the nationalized and planned economy based on state property remained.* As Trotsky explained in a 1939 article, "the working class of Russia accomplished the greatest overturn of property relations in history," and that conquest was still standing, despite the bureaucracy.

Trotsky analyzed this degeneration of the Soviet workers state, and he helped us integrate the political tasks flowing from it into a political strategy for all three sectors of the world revolution. This irreplaceable contribution to Marxist theory and to our program is contained in Trotsky's books *The Revolution Betrayed, In Defense of Marxism,* the Transitional Program, and other articles and polemics now available in English.

Today, more than forty years after Trotsky's death, a workers state, the dictatorship of the proletariat, still exists in the Soviet Union. World imperialism has not only failed to achieve a counterrevolutionary restoration of capitalism in the Soviet Union—a fact of enormous historical importance we tend to take for granted—but now confronts thirteen more workers states and several others on the way in Central America, the Caribbean, and Indochina.

State property, planning, the monopoly of foreign trade—these economic conquests of the Soviet working class have not been rolled back. This is true despite the fact that for more than half a century the Soviet Union has been *governed* by a privileged bureaucratic caste, a crystallized *petty-bourgeois* social layer with interests alien to urban and rural toilers.

This bureaucracy, however, is not the ruling class. The bureaucracy is not necessary to the development of these new property forms, and is in fact a parasitic obstacle to their full development. Nonetheless, in order to ensure its own survival and the survival of the trough from which it feeds its privileged habits of consumption, the bureaucratic caste must defend state property, although it does so inadequately and with counterrevolutionary methods that weaken and, in the last analysis, risk the destruction of the workers state. The political revolution to remove this parasitic caste is an essential part of the proletariat's defense of state property in the countries where capitalism has been abolished, and of the extension of the world socialist revolution.

Trotsky summed up the importance of this question for proletarian revolutionists in his 1939 article "The USSR in War," contained in *In Defense of Marxism,* a polemic against the petty-bourgeois opposition in the SWP.

> We must formulate our slogans in such a way that the workers see clearly just what we are defending in the USSR (state property and planned economy), and against whom (the parasitic bureaucracy and its Comintern). We must not lose sight for a single moment of the fact that the question of overthrowing the Soviet bureaucracy is for us subordinate to the question of preserving state property in the means of production in the USSR; that the question of preserving state property in the means of production in the USSR is subordinate for us to the question of the world proletarian revolution ([New York: Pathfinder Press, 1942, 1995], p. 73 [2014 printing]).

This succinct statement by Trotsky remains the fundamental strategic framework of our political orientation toward the deformed and degenerated workers states—and toward the political actions of their governments. Trotsky's first point is that the political revolution to overthrow the bureaucracy is for us a subordinate task to defense of state property. His second point is that assuring the survival of any particular workers state is a subordinate task to the extension of workers states on a world scale, as the part is subordinate to the whole.

This latter idea, by the way, is at the heart of the Cubans' proletarian internationalist perspective, as well. As Fidel explained in his report to the first congress of the Cuban CP in 1975: "The starting point of Cuban foreign policy . . . is the subordination of Cuban positions to the international needs of the struggle for socialism and for the national liberation of the peoples." And that is what they *do,* not just what they *say.*

Workers state

In coming to grips with the degeneration of the Soviet Union and the consequent political tasks, Trotsky had to sharpen up Marxist thinking about the criteria that define the class character of the state. This wasn't the product of idle theorizing, but of the need to grapple with new developments in the world class struggle in order to most effectively chart a course for the working class in that struggle. As a result, our movement in the 1930s developed an understanding of the dictatorship of the proletariat that we still use today—that is, a *workers state,* a state based on state property, planning, and a monopoly of foreign trade. We have defended this position against all comers in our movement who, under social patriotic pressures from bourgeois opinion, began to retreat from defense of these proletarian conquests, using the theoretical justification that the Soviet Union was no longer a workers state, but instead "state capitalist," "bureaucratic collectivist," or whatever.

For example, early heat lightning of the fight that would lead in 1940 to our split with the petty-bourgeois opposition appeared three years earlier during our entry into the Socialist Party. At that time, James Burnham and Joseph Carter—members of the Trotskyist left wing of the SP—made known their opinion that although "the economic structure as established by the October revolution remains basically unchanged," the Soviet Union did not remain a workers state "in the traditional sense given to this term by Marxism." It was, instead, neither a workers state nor a bourgeois state.

Trotsky answered Burnham and Carter in a November 1937 article entitled, "Not a Workers' and Not a Bourgeois State?".

> The class nature of the state is determined not by its political forms but by its social content; i.e., by the character of the forms of property and productive relations which the given state guards and defends.

Burnham and Carter had raised precisely the confusion over the dictatorship of the proletariat that we're discussing here. "The concept of the dictatorship of the proletariat is not primarily an economic but predominantly political category," they wrote. "All forms, organs, and institutions of the class rule of the proletariat are now destroyed, which is to say that the class rule of the proletariat is destroyed."

Trotsky replied:

> Of course, the dictatorship of the proletariat is not only "predominantly" but wholly and fully a "political category." However, this very politics is only concentrated economics. The domination of the Social Democracy in the state and in the soviets (Germany 1918–19) had nothing in common with the dictatorship of the proletariat inasmuch as it left bourgeois property inviolable. But the regime which guards the expropriated and nationalized property from the imperialists is, independent of political forms, the dictatorship of the proletariat (*Writings of Leon Trotsky (1937–38)* [New York: Pathfinder Press, 1970, 1976], pp. 75, 76 [2012 printing]).

Although a petty-bourgeois caste governs, the proletarian dictatorship remains. "So long as the forms of property that have been created by the October Revolution are not overthrown, the proletariat remains the ruling class," Trotsky explained in another polemic, this one written in 1933 and entitled, "The Class Nature of the Soviet State."

Trotsky's analysis also led him to look back with greater precision on the transitional government brought into being by the October Revolution. It was clear that the proletarian property relations that defined the continued existence of a workers state in the Soviet Union in the 1930s had not come into existence immediately following the October revolution. That took some time and leadership.

In his 1933 article "The Class Nature of the Soviet State," Trotsky wrote:

> Not only up to the Brest Litovsk peace but even up to autumn of 1918, the social content of the revolution was restricted to a petty-bourgeois agrarian overturn and workers' control over production. This means that the revolution in its actions had not yet passed the boundaries of bourgeois society. During the first period, soldiers' soviets ruled side by side with workers' soviets, and often elbowed them aside. Only toward the autumn of 1918 did the petty-bourgeois soldier-agrarian elemental wave recede a little to its shores, and the workers went forward with the nationalization of the means of production.
>
> Only from this time can one speak of the inception of a real dictatorship of the proletariat. But even here it is necessary to make large reservations.
>
> During those initial years, the dictatorship was geographically confined to the old Moscow principality and was compelled to wage a three-years' war along all the radii from Moscow to the periphery. This means that up to 1921, precisely up to the NEP, that is, what went on was still the struggle to establish the dictatorship of the proletariat upon the national scale.
>
> And since, in the opinion of the pseudo-Marxist philistines, the dictatorship had disappeared with the beginning of the NEP, then it means that, in general, it had never existed. To these gentlemen the dictatorship of the proletariat is simply an imponderable concept,

an ideal norm not to be realized upon our sinful planet (*Writings of Leon Trotsky (1933–34)* [New York: Pathfinder Press, 1972, 1975], p. 140 [2011 printing]).

No permanent détente

Despite the Stalinist bureaucracy's constant search for diplomatic deals with Washington to preserve the status quo at the expense of the world revolution, there can be no permanent détente. World imperialism cannot tolerate the existence of these new property forms. And they are determined not only to prevent their extension, but ultimately to roll them back in the fourteen countries where they currently exist. That is the long-range goal of the imperialists.

Trotsky explained this, too, in his 1937 polemic against Carter and Burnham: "In spite of all the efforts on the part of the Moscow clique to demonstrate its conservative reliability," Trotsky said, "world imperialism does not trust" it. Isolated counterrevolutionary services by the bureaucracy are not enough, Trotsky explained. Imperialism "needs a complete counterrevolution in the relations of property and the opening of the Russian market. So long as this is not the case, the bourgeoisie considers the Soviet state hostile to it. And it is right" (*Writings of Leon Trotsky (1937–38)*, p. 87).

With the existence of thirteen more workers states and several workers and farmers governments today, world capitalism is even more driven in this counterrevolutionary direction than when Trotsky wrote these words.

U.S. imperialism, in particular, cannot tolerate the extension of the socialist revolution *in this hemisphere,* especially under revolutionary leaderships such as the Cuban CP, FSLN, and New Jewel Movement. Washington is dead serious in its aim to overthrow the Grenadian and Nicaraguan workers and farmers governments, to prevent a successful outcome to the spread of the revolution to El Salvador, Guatemala, and elsewhere in Central America and the Caribbean, and—as U.S. officials themselves put it—to "go to the root," that is, revolutionary Cuba.

Trotsky also taught us that it is the counterrevolutionary pressure from imperialism against the Soviet workers state that is responsible for the bureaucracy's grip on political power over the workers. Ultimately, Trotsky said, the only road forward is the extension of the world socialist revolution.

One can with full justification say that the proletariat, *ruling* in one backward and isolated country, still remains an *oppressed* class [he said in his polemic with Burnham and Carter]. The source of oppression is world imperialism; the mechanism of transmission of the oppression—the bureaucracy.

If in the words "a ruling and at the same time an oppressed class" there is a contradiction, then it flows not from the mistakes of thought but from the contradiction in the very situation of the USSR. It is precisely because of this that we reject the theory of socialism in one country (*Ibid.*, p. 88).

This proletarian internationalist perspective is at the heart of any Marxist approach to the struggle against bureaucratization in any workers state—whether in one such as Cuba under revolutionary leadership, or in those where a bureaucratic caste has fastened a stronghold.

It is not accidental, in my opinion, that unclarity on the workers and farmers government has been accompanied in our world movement by growing weaknesses in understanding the importance of state property and why our movement stands for its unconditional defense against imperialism. The establishment of the Soviet workers state led to a qualitative shift in the world relationship of class forces to the benefit of workers and farmers around the world, a shift that was reinforced by the victory of the Soviet Union in World War II.

If it weren't for the existence of the Soviet workers state, the Chinese revolution could not have been victorious and could not have survived. Vietnam could not have won. The Cuban revolution could not have conquered and continued to advance and aid the revolution in Latin America and other countries. Nor could the Nicaraguan and Grenadian revolutions, whose victory and survival depend on the continuing revolutionary vitality of the Cuban workers state.

Because the imperialists have been unable to crush the Soviet workers state, the workers and farmers have been able to overturn capitalism in more than a dozen other countries over the past forty years. In that sense, state property does tend to expand beyond its own borders. It exercises a powerful attraction on the oppressed and exploited toilers worldwide. Thus, our defense of state property against imperialist counterrevolution is central to our proletarian internationalist goal of extending the world socialist revolution.

Moreover, as Trotsky explained, it is along this road that the workers and farmers of the Soviet Union and other bureaucratized workers states will be able to throw off the privilege-seeking parasites who have temporarily hobbled their march toward socialism.

The transition is key

So, the degeneration of the Soviet workers state provides another reason—one that Marx, Engels, Lenin, and the Comintern in its Leninist period could not possibly have foreseen—why it is so important when there is a workers and farmers government to keep our eyes on the *transition* from the dominance of inherited bourgeois economic relations to their expropriation and thus the establishment and consolidation of proletarian economic relations—i.e., state property.

Until that transition is completed, a workers state does not yet exist. Once it has been completed, this transition is a much more durable conquest of the workers—a conquest that will permanently face the threat of capitalist counterrevolution from beyond its borders, but one that can only be overthrown by a capitalist counterrevolution. This will remain true until the last stronghold of imperialism has been defeated by the world proletarian revolution.

As Trotsky explained in *The Revolution Betrayed* with regard to the Soviet Union:

> As a conscious political force the bureaucracy has betrayed the revolution. But a victorious revolution is fortunately not only a program and a banner, not only political institutions, but also a system of social relations. To betray it is not enough. You have to overthrow it.
>
> The October revolution has been betrayed by the ruling stratum, but not yet overthrown. It has a great power of resistance, coinciding with the established property relations, with the living force of the proletariat, the consciousness of its best elements, the impasse of world capitalism, and the inevitability of world revolution ([New York: Pathfinder Press, 1937, 1972], p. 261 [2013 printing]).

This is why we insist that it is neither sufficient nor correct to say that the workers and farmers government is merely a popular designation or synonym for the dictatorship of the proletariat. That misses this all-important *transition,* which culminates in another qualitative turning point in any anti-capitalist revolution—the expropriation of the exploiting class and the establishment of state property. Only when this has been accomplished do we have a workers state—which *is* fundamentally a popular designation or synonym for the dictatorship of the proletariat.

Nor is this the first time we've had to grapple with this question as a result of working out our theory of the workers and farmers government. Joe dealt with it back in 1961 in a polemic against Tim Wohlforth, James Robertson, and Shane Mage entitled "What the Discussion on Cuba Is About" (Joseph Hansen, *Dynamics of the Cuban Revolution* [New York: Pathfinder Press, 1978], pp. 130–72 [2010 printing]).

> *State and Revolution,* excellent as it is in bringing together the teachings of Marx and Engels as the foundation for everything that followed, does not contain the final word on how to determine the character of a state. It lacks the refinements introduced as a result of subsequent experience and subsequent development of Marxist theory.
>
> Written in August–September 1917, it lacks in particular a consideration of what the Bolsheviks discovered in life after they came to power. It tells us nothing, for instance, about the experience of the Bolsheviks in facing the contradiction between government and state and resolving it. Not a word appears in it about the contradiction between government and state in the case of degeneration of workers' power.
>
> We need not lament this limitation in Lenin's famous pamphlet. Trotsky brought the criteria presented in *State and Revolution* up

to date as he followed the development of the first workers' state. In fact everything Trotsky wrote in relation to the character of the workers' state is built on the foundation of those teachings. *Built on.*

In that same polemical article, Joe explained how we could now look back at the stages of the Russian revolution with greater precision—on the basis of the experience of the Cuban revolution and our evaluation of it in light of the workers and farmers government discussions at the Fourth Comintern congress.

> . . . if Russia was called a workers' state in 1917, it was because everyone knew that the contradiction between the government power and the capitalist state it took over would be resolved by the establishment of a new state structure conforming to the Bolshevik program.
>
> Let us not fail to observe, however, that the promissory note did not in itself wipe out the contradiction [Joe cautioned]. This was only resolved in life itself, as Trotsky was to point out when he came to study the contradiction between the petty-bourgeois Stalinist power and the workers' state it rested on.

So, we can call the workers and farmers government a *bridge to* the dictatorship of the proletariat, the *antechamber of* the dictatorship of the proletariat, or the *first stage of* the dictatorship of the proletariat. They all amount to about the same thing.

But it's wrong to say that it can only be a synonym for the dictatorship of the proletariat. Instead, our transitional conception of the workers and farmers government correctly focuses on the *political and economic tasks* that the proletariat must lead its toiling allies in carrying out, even after the revolution has triumphed, dual power has been resolved, and the old bourgeois state apparatus and army have been replaced. Our transitional approach provides a guide to the way *toward* the consolidation of a workers state, which is the most durable conquest of the proletariat in any anticapitalist revolution.

We don't use the term "dictatorship of the proletariat" much anyway, for obvious reasons. Working people in the twentieth century have had enough "dictatorship"; the term is hardly the most useful tool in getting across the point that communists need to explain.

We normally talk about workers and farmers governments and workers states. Establishing the first one helps you get the other. And once you've got that, you can begin doing what they've been doing in Cuba for twenty years and what they're trying to do in Nicaragua and Grenada today. That's simple and accurate. That's what we're talking about.

V. FROM CHINA TO CUBA, NICARAGUA AND GRENADA

World War II was followed by two great conquests of the world proletariat—the overturn of capitalist property relations in Eastern Europe, including the Yugoslav revolution, and, most decisive for the world relationship of class forces, the revolution in China—the most populous country in the world. The fact that these revolutions took place under Stalinist leadership *was* unexpected by us, and it posed a gigantic question to come to grips with.

The problem in Eastern Europe was easier. It was one of recognizing, after the fact, that states that could only be described as having the economic structure of workers states similar to the Soviet Union had come into being. Trotsky's writings in *In Defense of Marxism* were especially helpful in shaping our views on this. We faced no major obstacle in coming to the political conclusion that these states were to be defended unconditionally against imperialism, even before the transition to the new economic relations was completed. A substantial majority of the Fourth International came to those positions and acted on them. There were complications in analyzing Eastern Europe and Yugoslavia, but we don't have time to go into that here.

China was a bigger question. There was no way of reading or interpreting the Transitional Program that would lead anyone to expect that the Stalinist Chinese Communist Party, which Trotsky characterized as a peasant party with a petty-bourgeois leadership by the early 1930s, would end up leading the Chinese toilers to power and then leading them in establishing a workers state, with all its implications for the world revolution.

There is no way to stretch either the Comintern resolutions or anything that Trotsky subsequently wrote that would have led us to believe that a workers state could be established without revolutionary Marxists gaining the dominant leadership in the transitional period. In fact, the Comintern and Trotsky had precluded such a development. A workers and farmers government was conceivable without a revolutionary proletarian majority, they believed, but *not* a successful transition to a workers state.

But the facts turned out differently. And the split that took place in the Fourth International at the end of 1953 unfortunately precluded a common discussion of the theoretical questions involved.

Both sides of the divided International subsequently came to the conclusion that a deformed workers state had been established in China. The SWP's resolution on this question, "The Third Chinese Revolution and Its Aftermath," was adopted by the National Committee in September 1955, and then by an SWP convention nearly two years later in mid-1957. There was broad agreement in the party leadership on the enormous significance of the revolution for the world class struggle and the political tasks of defending China that flowed from this. The party's record in the Korean war and in defense of the Vietnamese revolution were exemplary chapters in our history.

Nonetheless, we didn't pretend to have put into place all the theoretical questions and problems posed by the Chinese revolution. Our eyes were still primarily focused on the question of the characteristics and limits of the Stalinist leadership that presided over the transition to a workers state in China and thus the limits to any *pattern* of revolution with that caliber of leadership. We did not focus on the question of what kind of transitional government existed in China and the role it played in establishing a workers state.

Joe was among the most dissatisfied with those unresolved questions and he abstained on the China resolution at the September 1955 National Committee plenum. He placed a short statement in the minutes stating, "I agree with the general political line as developed in the resolution [on China] but I disagree on characterization of China as a deformed workers state." But his alternatives to the adopted resolution didn't satisfy either him or his collaborators in the central leadership of the party.

Joe had become acquainted with the Fourth Comintern Congress discussions on the workers and farmers government through French-language editions, and he had gotten hold of an old British translation of some of the resolutions and theses adopted at the Fourth Congress. Then, in late 1956, a book of English translations of major excerpts from the main documents adopted at the first four congresses of the Comintern was published by Oxford University Press (*The Communist International, 1919–1943, documents,* Volume 1, 1919–1922, selected and edited by Jane DeGras). Joe was urged to work up some classes on it in California, where he and Reba [Hansen] had moved in the summer of 1957 to help Jim Cannon with a book and to continue editing the *ISR* from out there.

In September, a little more than a year before the Cuban victory, Joe presented those classes to a summer school in California; he called the series, "The Four Forgotten Congresses of Lenin's Time." That seems like a jarring title to us today, when so many documents of the early Comintern are now available in English. But until the end of the 1950s, their general inaccessibility in English had virtually made them "four forgotten congresses" for most revolutionists here.

Joe wrote a review of the new book in the *ISR* in 1958, and ran an advertisement encouraging readers to order it, along with Trotsky's *First Five Years of the Communist International.*

Then in January 1959 something unexpected came along that could be studied along with the Comintern discussions—the Cuban revolution. We threw ourselves head over heels into defense of the Cuban revolution—our revolution—in the early 1960s. Moreover, that revolution also provided the occasion for the party to make some new progress on all the theoretical questions posed by the postwar events—including the Chinese revolution.

Here Joe's growing interest in and study of the Comintern documents really bore fruit. These documents drafted by the Bolsheviks helped anchor our search for a solution to the unresolved theoretical problems in the transitional approach that permeated the resolutions and reports adopted by the Comintern while it was under Bolshevik leadership. We followed the development of the Cuban revolution very closely and concretely. And it became clear that these Comintern discussions of Lenin's time offered a needed theoretical starting point to understand the dynamics of the Cuban revolution. Our approach to the Cuban workers and farmers government, in turn, helped us look back at the evolution of the Chinese revolution—and later Yugoslavia and Eastern Europe—with greater theoretical clarity. Joe summarized some of our major conclusions in a report adopted by the 1969 party convention. [See Appendix VIII].

Cuban revolution

In China, a Stalinist leadership had stood at the head of the mass revolutionary battles that led to the establishment of a workers and farmers government, which ended up carrying through the expropriations of the bourgeoisie. In a situation of an extremely decayed, discredited bourgeois regime, and under extraordinary pressures of war and the threat of counterrevolution by imperialism and domestic reaction, this leadership proved adequate to mobilize the workers to overturn capitalism and establish a workers state. That's a fact.

But a deeper question was involved, because we did not for even a minute draw the conclusion that this caliber of leadership could advance the world socialist revolution—not only in countries such as the United States and in France, but even as a general pattern in semicolonial countries with weaker regimes. We did not for even a minute call into question the Leninist strategy of party building, the transitional method, or the need for mass proletarian revolutionary parties and a world party.

This is why the Cuban revolution helped break the theoretical logjam we had been in since the early 1950s. It was here, for the first time since the Bolsheviks, that we saw the power of a workers and farmers government *headed by a revolutionary leadership.* Here we could see what a revolutionary leadership could do with this powerful tool. The political course of the Cuban workers and farmers government, at home and abroad, was something revolutionists around the world could use to inspire and educate working people in a new way.

The program of the July 26 Movement was a radical democratic program, one that the Cuban revolutionists used to mobilize growing sectors of both the rural toilers—the sugar workers, the peasants, the townspeople—and the urban toilers to bring down the U.S.-backed dictatorship. When the revolution triumphed, the initial provisional revolutionary government was a coalition of the revolutionary forces and prominent bourgeois opposition figures.

From inside and outside the government, the July 26 Movement under Castro's leadership continued to mobilize the Cuban workers and peasants to implement the program they had been struggling for. The people were armed; the revolutionary army was strengthened and militias were organized. Fidel's famous 1953 speech, *History Will Absolve Me,* served as the initial program. Radical measures, including transitional measures, were fought for and implemented—agrarian reform, anti-imperialist actions, measures to improve the conditions of the workers and expand their control over production, as well as the elimination of government corruption, gambling, prostitution, and racist discrimination.

As the revolution marched forward, class differentiations within the governmental coalition sharpened. These class pressures forced the resignation of key bourgeois representatives in the government. By mid-1959, a workers and farmers government had been established. This revolutionary government continued to promote the organization and mobilization of the toilers which took place more and more under the leadership of the working class. The process culminated in the establishment of the first workers state in the Americas in the summer and fall of 1960, as massive popular expropriations of imperialist and native capital took place.

We could now see in a new and concrete way the power of *advocating* a workers and farmers government, and fighting for the kind of revolutionary leadership that was mobilizing and organizing the Cuban workers and peasants.

The Cuban revolution also led us to a more accurate appreciation of the tactic of guerrilla warfare as part of worker and peasant revolutions in the colonial and semicolonial world. In the document "For Early Reunification" drafted by the SWP Political Committee in March 1963 for consideration by the divided forces of the world Trotskyist movement, we included the following as one of sixteen proposed points of common outlook:

> Along the road of a revolution beginning with simple democratic demands and ending in the rupture of capitalist property relations, guerrilla warfare conducted by landless peasant and semiproletarian forces, under a leadership that becomes committed to carrying the revolution through to a conclusion, can play a decisive role in undermining and precipitating the downfall of a colonial or semicolonial power. This is one of the main lessons to be drawn from experience since the Second World War. It must be consciously incorporated into the strategy of building revolutionary Marxist parties in colonial countries (*Dynamics of World Revolution Today* [New York: Pathfinder Press, 1974], p. 19).

The Cuban revolution and its leadership showed in practice that the transitional program, our program, is not some way to trick the toiling allies of the working class. The call for a workers and farmers government is not some public relations promise to the peasants to get them to accept working-class leadership, in order that they can then be "dictated to" by the proletariat after the revolution triumphs.

No, such a government seeks to mobilize and organize both the workers and working farmers against their common exploiters, in order to carry out the concrete transitional measures necessary to resolve the problems facing the nation and to defend, consolidate, and advance the social revolution. How fast these measures are implemented and how they are done will be determined together by the workers and farmers, depending on the concrete course of the class struggle. As the inevitable class differentiations occur in the countryside and in the cities, and with proper leadership, the economic foundations of a workers state will be established and consolidated.

The revolutionary Cuban government, from the outset right through to today, has proved in action to the farmers—as Trotsky put it—"This is also your government."

In Cuba we have seen a living example of the working class marching at the head of all its oppressed and exploited allies. In this sense, the Cuban revolution became a model for the Socialist Workers Party in a way that the Chinese revolution could not. Because for the first time since the Russian revolution, this process was being led by a genuinely revolutionary leadership—one not deeply marred by training in the school of Stalinism. And by a leadership that proletarianized itself and became more communist in the course of deepening the revolution—and has continued to do so to this day.

We responded to the Cuban revolution by sending Joe and Farrell [Dobbs] down to Cuba in 1960 to take a firsthand look and bring back the truth to the American people to answer the barrage of lies in the big-business press. Joe wrote a series of *Militant* articles on the basis of the trip, and Farrell, who was our 1960 presidential candidate at the time, spoke about it at meetings across the country. So the Cuban revolution became a model for us in the early 1960s, in the same way that the Nicaraguan and Grenadian revolutions are today.

We point to their accomplishments and their leaderships as a way to educate and inspire workers and farmers in the U.S. today. We say we are striving to build a similar kind of party, one that can lead the working class and its allies in this country to establish the same kind of government, a workers and farmers government. We want to implement the same kinds of economic and social measures in the interests of the vast majority of working people, and we point out what could be accomplished with the massive wealth and productive capacity of the United States.

We say we want to follow the Cuban road, the Nicaraguan road, the Grenadian road.

There was agreement among the forces that came together to reunify the Fourth International in the early 1960s on welcoming the Cuban revolution, recognizing the revolutionary character of its leadership, and acknowledging the existence of a workers state there by the end of 1960. But the Fourth International has never adopted a statement by the United Secretariat, the International Executive Committee, or a World Congress recognizing that the Cubans established a workers and farmers government, and that this was the key powerful tool for leading the workers toward the expropriation of the bourgeoisie and the consolidation of the Cuban workers state. Nor has the earlier view of the revolutionary character of the leadership been maintained.

At one time, there seemed to be a common recognition that a workers and farmers government had come into existence in Algeria, as I've already mentioned. But that agreement began unwinding a couple of years later, as a disagreement developed over whether or not the 1964 resolution should be the basis of the 1969 United Secretariat resolution on Algeria.

So, what we're dealing with in this report is not the Fourth International's record on the workers and farmers government. We're dealing with the *SWP*'s evolution on this question, and we are today discussing a further clarification and deepening of our understanding on this question.

There remain big political differences over this question in the Fourth International today. The difference over the workers and farmers government in Nicaragua was, in my opinion, the sharpest one at the 1979 World Congress. That debate was preceded by a growing divergence over the caliber of the FSLN leadership and the correct approach to it—a divergence that took its most heated form over the different reactions within the International to the Morenistas' Simón Bolívar Brigade adventure.

The emerging differences over Nicaragua led to calling the only caucus meeting in an elected body of the Fourth International between the time of the dissolution of the factions in the Fourth International in 1977 and the World Congress in 1979. That occurred at a United Secretariat meeting shortly before the World Congress, when Barry [Sheppard] and Caroline [Lund] reported the PC discussion where we had arrived at our assessment that there was a workers and farmers government in Nicaragua under a revolutionary leadership. Other comrades on the Secretariat immediately called a caucus meeting of those who did not agree with this position. That is how important they considered this political question. The same differences exist in relation to Grenada and remain unresolved.

I point this out in order to avoid giving an inadvertent impression that there is agreement on the workers and farmers government in the International. There most certainly is not. But in this report we're not dealing with the Fourth International's discussions on this question.

VI. IN THE UNITED STATES TODAY

What about today? What are we, the SWP, going to say and do about the governmental question in the United States? What should our slogan be?

We haven't just pulled this question out of thin air. It has been posed in life for the party. It is one of the results of the turn. It is part of coming out of the quarter of a century of semisectarian existence imposed on the party by the postwar reaction, economic expansion, and political retreat of our class. It is part of dealing with the living reality of this country, of the *whole* country and its internal class relations. It is part of explaining to working people how socialists propose to solve the real problems facing the country—including how to feed ourselves.

Other political currents are thinking about these questions, too. The NBIPP included a section on Black farmers and agriculture in the program adopted at its founding conference last summer. The union movement is often faced with the question of these allies.

Through our involvement in the class struggle in larger and more diverse areas of the country than ever before, we're obliged to think about and learn more about the connection between farm workers, farmers, various layers of farmers, and the large number of semiproletarian farmers who work alongside us in many factories around the country. We're learning concretely that rural America is not just farmers and their families, but a network of farmers, townspeople, workers, and small businessmen that differs from one section of the country to another. We're paying more attention to the connection of agriculture in the United States to interimperialist competition, world economic relations, and the world class struggle.

We've been able to make progress in understanding this because of the turn. Right after we began the turn in 1978, we asked John Staggs and Syd Stapleton to prepare some classes on farmers in the United States for the national education conference that summer. Osborne Hart and other comrades subsequently began to follow this question, as well, and helped us become more informed about it. At our spring plenum in 1979, Doug Jenness gave a report on agriculture in the U.S. that was adopted by the National Committee, and printed along with other material Doug collected in an Education for Socialists bulletin. [See *Marxism and the Working Farmer.*]

The party has also been going through political experiences in parts of the country where we haven't had branches in a long time—experiences in Nebraska, Kansas, Missouri, in the coal fields, and in the South and Southwest.

All these things made us think more deeply about the totality of American society and its working population.

Town and country

The question of the government is today posed more sharply for both workers and farmers in this country than it has been since the Great Depression and preparations for World War II. Socialists need to be able to explain what kind of government is needed, how it can be fought for, and what alliance of classes can achieve it. We need to explain why a workers and farmers government is a life-and-death question. How such a government could shelter, clothe, transport, and feed society. How it could unleash the energies of the toilers to transform life in this country and help raise the living conditions of people throughout the world. That requires more than a couple of slogans, even correct ones.

Far from being pushed farther apart today by the deepening social crisis of American capitalism and the rulers' growing use of their military power, city and country are being pulled closer together.

Urban and rural working people are suffering many of the same brutal blows. What capitalism has in store is depression and war. The alternatives we all face are socialism or barbarism—or worse.

Workers and farmers confront the same staggering interest rates; city workers trying to get a car or home loan have a better idea today than for a long time what it means to be a debt slave in the countryside. Both workers and farmers are burdened by rising fuel, water, and electricity rates; hence both have a stake in the fight to nationalize the utilities and Big Oil and place them under the control of working people. Both have an interest in fighting the trade barriers and embargos being thrown up by the government for economic and political reasons that have nothing to do with the interests of working people. The dangers of nuclear power and destruction of the environment are issues of concern to both city and country.

War and the reality of how Washington might use its mammoth nuclear stockpile are also questions that face both workers and farmers. Neither want the new Vietnams that have already begun in Central America. And Ground Zero will be everywhere.

Andrew [Pulley] and Matilde [Zimmermann] tell me that when these kinds of questions came up during talks and interviews in the 1980 election campaign, they frequently responded by talking, among other things, about the need for a workers and farmers government. This slogan has also begun cropping up in election campaign platforms and leaflets, and for the same reasons. It's a response to changes taking place in this country, as well as changes in where and with whom we're doing political work today.

It is vital for the SWP, as the nucleus of a mass revolutionary party of the American working class, to provide correct answers to these questions of the countryside and the city, the workers and the farmers, and what kind of government can address their problems. Because if *we* don't find ways of integrating solutions to these problems into our propaganda work and our political perspectives, *then others will fill the vacuum with wrong and dangerous answers.*

As the capitalist offensive bears down with increasing ferocity on the toilers, there will be more and more varieties of what Marxists have called "fools' socialism"—"neopopulism," schemes to get rid of interest rates, or abolish gold, or scapegoat Jews, or whatever. All this starts mounting as the crisis deepens. Every single petty-bourgeois and reactionary current of thought that Marx, Engels, Lenin, and Trotsky had to combat will reemerge as a panacea, raised by some section of misleaders in the workers movement and by layers of the petty bourgeoisie. We would be hopelessly naive to think otherwise.

That's why it's important for the proletarian party to develop what Trotsky in the 1930s called "a program of transitional demands for the 'middle classes,'" especially for the workers' allies among the other exploited producers.

Any time that the vanguard of the working class does *not* give its allies clear, concrete answers that point a way out of the crisis of capitalism and does not orient the labor movement toward collaborating with its allies, then this default opens the road to defeat for our class. This is not merely or even primarily a matter of the interests of the workers' allies, either. The main commitment is to the working class. The workers have a right to ask, "How are we going to be fed under this new government that you socialists advocate?" If we're serious about governing, that's an elementary question we have to be ready to answer.

The 1967 change in our slogan

We did not discuss this question in the late 1960s when we changed our slogan from "For a workers and farmers government," which had been our transitional governmental slogan since 1938, to "For a workers government."

We made that shift at our 1967 convention. There was no special report on this change, however. And there was virtually no discussion prior to it. The proposal to make such a change had been placed before the National Committee almost three years earlier in a letter from Detroit by George Breitman. [See Appendix IX.] Other than this original proposal, all that appeared in the internal discussion bulletin on this matter was a two-and-a-half page motion by the Political Committee making the proposal shortly before the 1967 convention, and an accompanying one-and-a-half page article providing some recent farm statistics that the PC had asked a comrade on the *Militant* staff to prepare.

The PC motion [see Appendix X] provided a summary of the role of the governmental slogan in our program, based on previous party documents. It correctly established that the general content of the two slogans is the same and approached them flexibly.

> Both are designed for common use as a bridge to the idea of a revolutionary government of the workers and their allies. This transitional concept was developed by the early Comintern then led by Lenin and Trotsky. Its purpose was to initiate mass consciousness of the need for class struggle politics, as against the social democratic line of political coalition with the capitalists, and to develop that consciousness to the logical revolutionary conclusions.

That is important to understand. For one thing, it means that we shouldn't be worried about the election campaign leaflets from around the country over the last couple of years that have already raised the proposal for a workers and farmers government. These is no *principled* difference between the two; used correctly, both are transitional slogans pointing to the need for a government of the workers and their allies.

The PC motion then went on to state that the "relative political weight" of the farmers, "as compared with other potential allies of the workers," was such today that they "no longer constitute an especially significant force to be singled out above all other potential allies of the working class" in our transitional governmental slogan.

Nearly three years went by after the first proposal from Detroit before we finally acted on the change in slogan. Most of the comrades on the Political Committee at the time weren't initially certain there was any good reason to change. There is a correct caution and conservatism in the party about making such changes. But more and more comrades became convinced over time that there was no reason *not* to alter the slogan, given the decline of the size of the farm population.

The Political Committee discussed and rejected several alternative ways suggested by Comrade Breitman of including various allies in the governmental slogan—"For a government based on, representing and acting for Black and white factory, farm and office workers"; "A Black and White Workers Government"; "A Factory, Farm and Office Workers Government"; and "A government of factory, farm and office workers and their allies."

The final result was the PC motion adopted by the 1967 convention. The resolution correctly did not include any of the proposed alternatives, and it clearly explained the transitional use of the governmental slogan. Thus, while our experience from the turn in the late 1970s and early 1980s convinces us that the decision to drop "farmers" from the slogan was wrong, the correction we're proposing here today is not a shift in our basic views. It is in harmony with our programmatic continuity on this question.

What was the political heart of the 1967 decision to change the slogan?

If you read the small amount of written discussion material on it, you can't come to any other conclusion but that it didn't have much to do with farmers at all. There was no serious discussion about rural America, the role of agriculture in the U.S. economy, the class structure of the countryside, land ownership, and so on. The fact that farmers comprise a smaller and smaller percentage of the U.S. population was the only argument brought to bear—literally the only one. This is a correct factual point, but it was not sufficient to justify the change.

So, that clearly wasn't the driving force behind the change.

What *was* involved was something important and progressive—the party's growing attention to the Black struggle. As those of us who were in the party at the time remember very well, we were trying to find every possible way to integrate ourselves into the rising struggle by Black America for liberation. We sought to use everything in our political arsenal, including our governmental slogan, to advance this perspective.

After some discussion, as I said, the idea of raising some alternative governmental slogan was rejected; no one moved that at the 1967 convention. But, as the PC resolution put it, there no longer seemed to be a reason why farmers should "be singled out above all other potential allies"; the Black population was the potential ally that this sentence was referring to in particular.

A wrong decision

So, the motivation was progressive, but the reasoning and decision were wrong. What was wrong with them?

The first thing was that the change was carried out in ignorance of farmers, U.S. agriculture, and the countryside. In 1967—with a few exceptions—we were not involved in any meaningful way in arenas of activity that brought us into contact with farmers. Through no fault of our own, we were not the kind of party that we are becoming today because of both the changes in U.S. and world politics and our turn to industry.

Today, fifteen years later, farmers are an even smaller percentage of the population than they were in 1967. That trend continues, reflecting the squeeze on working farmers under capitalism. But our experiences and further thinking have made clear that there's a lot more to the question than that. We've had more contact with farmers and rural America. We have been involved in and have written about protests by farmers. We have learned about the issues that affect them. We've found out about the overlap between the working class and the farm population in many parts of the country. Most important of all were the experiences of our industrial fractions, as well as thinking out how our election campaigns can relate to the entire working class and its allies, including farmers and other working people in rural areas.

It's *this* that has changed our understanding of the farm question since the end of the 1970s.

In 1967, however, we didn't yet see that the decisive question is not the size of the farm population. Our eyes weren't on the character of exploitation in agriculture and the economic, social, and political weight of American farm production in this country and its interconnection to world politics.

These factors make the political significance of the small size of the farm population in this country different from its significance in Britain. Trotsky explained that in Britain farmers are numerically and *economically* negligible. The food question there is more one of imports than one related primarily to British agricultural production and exports. Let's leave aside how that has evolved since Trotsky's death, and what it may or may not mean about the correct governmental slogan in Britain. That's up to the British comrades to discuss and decide. Given the existence of the mass Labor Party there, some way of calling for a Labor government is the concrete form often taken by the slogan either for a workers or workers and farmers government. (After World War II, the comrades in Europe made use of the workers and farmers government slogan extensively in 1945–46. I won't go into that, but you can read about it in a 1946 article by Michel Pablo which is included in the Education for Socialists bulletin on *The Workers and Farmers Government.*)

My point here is only that if there is one thing

that is *not* economically negligible in the United States, it is American agriculture. American farms and the farming population are *not* negligible in this country—economically, socially, or politically. And they don't play a negligible role in the world either, nor are they nor will they be negligible to the world revolution.

To the contrary, American farming is a weighty economic factor, here and around the world, as the report by Doug [Jenness] on "Agriculture and Farmers in the United States" that we adopted earlier in the plenum amply demonstrated. [See *New International* no. 4.]

But we didn't figure this out by finding some better facts and figures. It took the turn and what came out of the turn. That's what enabled us to take another look at this question and make the necessary shift that we're proposing the National Committee adopt at this plenum.

We're fortunate that we began preparing for this discussion early on in the turn. As I mentioned earlier, at our May 1979 plenum the NC adopted a report given by Doug for the PC on "American Agriculture and the Working Farmer." The relation between our governmental slogan and what we were already learning about the farm question came up under the discussion on that report.

> I think this is one of the questions we'll want to consider further in the course of our discussion on the farm question [Doug said in his summary].
>
> We should bear in mind that we did not change our slogan [in 1967] as the result of a thorough discussion about American farmers—their social weight and role in the economy. Rather, the revision grew out of a discussion of how we could include other weighty allies of the working class, especially oppressed national minorities, in our governmental slogan. When we didn't come up with a satisfactory alternative on this, we decided to drop farmers from our governmental demand in consideration of the decrease in the numerical size of the farm population.
>
> Now we can rethink this question. But it's better to have our discussion and establish our position on American agriculture first. Then we'll be on firmer ground in the decision on our governmental slogan. [See *Marxism and the Working Farmer,* an Education for Socialists publication.]

Based on the experiences and knowledge we've accumulated since that time, pulled together in *Militant* articles and in Doug's report adopted at this plenum, we are certainly on "firmer ground" in making the decision that we are proposing for adoption here today.

Allies of the working class

There was another problem with the 1967 decision, too. It failed to make any distinctions among the various allies of the working class: those who are primarily an especially oppressed layer of the working class; those who are oppressed and superexploited because of their nationality, sex, or age; those who are not part of the class of wage workers but are nonetheless producers brutally exploited by industrial, banking, commercial, and landholding capital.

Let me cite one example of how incorrectly we were thinking about this question for a while—a paragraph from the introduction to our book on the Transitional Program. This is what it says:

> We haven't worked out a program for the American peasants, because we don't have so many. There are only about 5 million farmers compared with 9.5 million college students in this country. Thus the student population, which is in motion, is more important than the farm population, politically speaking. Indeed several years ago we amended our transitional demand for a new regime from "a workers' and farmers' government" to a "workers' government." In some areas—for example, in Minnesota—we may require an extensive set of demands for the small farmers in the future as we have had them in the past. However, this is not now a central concern for us on a national scale (p. 88).

We read this today and we're horrified. But how many of us read this when the book first came out in 1973 and weren't horrified at all? Why? It's simple—because we can't always leap ahead and anticipate how life will develop. The paragraph went right by all of us because of where we were doing political work at that time, with whom, and what we were thinking about as a result.

In changing our governmental slogan back to "for a workers and farmers government," we can also untangle confusion between two tasks facing revolutionists in the United States: first, finding ways to explain the combined character of the American socialist revolution and to advance a program for Black rights, for women's rights, for the rights of youth; and second, finding ways to use our transitional program to build a bridge to an alliance with *other exploited producers* in this country, especially farmers.

We're accustomed to thinking of the workers' allies primarily as *oppressed* layers of the population.

Blacks and Chicanos, for example, are oppressed *nationalities*. This national oppression makes Black and Chicano workers subject to superexploitation by the capitalists. But as nationalities, Blacks and Latinos are *oppressed*. Their fight for national self-determination is an unsolved democratic task, although we know that only a successful proletarian revolution can open the door to its solution.

Women are an *oppressed sex.* Once again, this special oppression makes women workers subject to superexploitation. But women as a sex are an *oppressed* layer of the population.

Revolutionary Marxists recognize that the fight against the oppression of minority nationalities and women is a form of the class struggle; all oppression under capitalism breeds resistance that are forms of the class struggle. The working class must advance a program that champions these struggles and provides concrete solutions to the problems facing all the oppressed. Given the proletarianization of the oppressed nationalities and, to a lesser but important degree, of women over the past several decades, the social weight of these layers of the population and the political centrality of the issues they raise have become even more pivotal to any successful strategy for making the American socialist revolution.

But working farmers are not an *oppressed* sector of the population, in any special way. As small independent commodity producers, they are *exploited.* Industrial, financial, commercial, and landed capital take a big, unpaid chunk out of the product of their labor through rent and interest payments and through monopoly-rigged pricing both on the buying and selling end. The exploitation of working farmers takes forms that are different from the wage slavery of the working class, but both workers and farmers are *exploited* by the capitalist class. (Of course, a certain number of wage slaves are also exploited farmers!) Both workers and farmers have a big stake in expropriating those exploiters. That's a strong material basis for the worker-farmer alliance we're determined to help forge in the fight for a socialist America.

That's where our governmental slogan comes in. We are fighting for a government of the exploited producers that can expropriate the exploiters, establish a workers state, and open the road to socialism and to ending all the effects of oppression. That's the transitional function of our governmental demand.

A year after our 1967 convention, the Canadian comrades decided *against* a proposal, occasioned by our decision, to drop their slogan for a workers and farmers government. Dick Fidler wrote a good article in the Canadian discussion bulletin that dealt with the role of the countryside in Canadian politics, the political role of farmers, the argument about the declining size of the farm population, and so on.

Of course, no one in the SWP proposed that we raise the slogan of a workers and students' government, or a workers and women's government, or—by the time of the convention—a Black and White workers government. But you can see the incorrect framework it was easy to slide into.

We weren't looking at the slogan as a way of presenting *a government of the producers who are exploited by the capitalists,* a government that can lead the popular masses forward and carry out a program to rid society of all forms of exploitation and oppression perpetuated by capitalism.

Vanguard role of Black workers

This mistake on our governmental slogan actually cut against the grain of our deepening understanding of the vanguard role of the Black struggle in the American socialist revolution. It didn't jibe with the vanguard role that we are convinced Black workers will play in the battle for socialism and for a proletarian party. We hadn't thought it through to the end.

To understand how we could have made such an error, we have to remember the class and social composition of the party in 1967, as well as our political experiences. We were already getting more involved in the struggles of Black Americans, but we still had a very small Black cadre. Not all of us were yet really looking at Black workers as a vanguard component of the working class *as a class*. We weren't sufficiently taking account of Black workers as makers and leaders of the American socialist revolution, disproportionate to their numbers. We couldn't jump over our own limitations at the time and confidently assume the role that worker-Bolsheviks who are Black would be playing fifteen years later in leading our party, leading our industrial fractions, leading our work in other areas.

In 1967 many of us were still thinking in terms of how the socialist movement could begin to appeal to the Black population. And that affected how we approached the governmental slogan. Many of us were concerned that the workers and farmers government slogan would be read by Blacks as either excluding them or giving a back seat to their struggle for self-determination.

Today we can see the error in this approach. We can see that the actual question is: How does a revolutionary proletarian party—more Black (and more Latino) in composition than the population as a whole—present a perspective of a transitional government to the workers and the workers' allies among other exploited producers? Unless we think that Black workers, who are in the vanguard on other political questions, have some peculiar inability to grasp the farm question, it makes no sense to drop farmers from our governmental slogan using the argument that Blacks are a larger and weightier ally.

Of course, there is no political question that vanguard workers who are Black and Latino are less capable of grasping than other workers. In fact, as the class struggle accelerates in this country, our transitional demands and our governmental slogan will more and more become the program of a class-struggle wing of the labor movement led by workers who, in comparison to the overall makeup of the population and even of the working class, are disproportionately Black and Latino.

Our approach to the Black struggle is built around its vanguard role in the American class struggle, the combined character of the American socialist revolution, the vanguard role of Black workers in transforming the labor movement, and unconditional support for the right of self-determination including separation. That is part of the bedrock of our party's political conquests reaffirmed over the past two decades.

We know that the capitalists profit from national oppression and that a government advancing the employers' interests will never recognize the right to self-determination. Far from being a reason *not* to advocate a workers and farmers government, our unconditional support to Black self-determination is just the opposite. Self-determination is *not* a governmental slogan. It is a fundamental democratic right that Blacks and Latinos can only achieve through the successful struggle to achieve a workers and farmers government.

At some stage in its development, I'm convinced that the NBIPP will discuss the perspective of this kind of government. As vanguard fighters for Black rights who recognize that capitalism and imperialism are the source of racism and national oppression, they will have to deal with the need to advance the perspective of a new kind of government representing workers and farmers, who have *no* interest in the oppression of Blacks or any other layer of the population.

VII. NOT BACK, BUT FORWARD

So, that's where we've come from on the governmental question and why we should decide here to adopt the slogan, "For a workers and farmers government." We're not going *back* to an old slogan; we're moving *forward* to adopting a slogan that we can apply and use concretely in the political situation our class faces today.

In adopting this slogan, it's important that we understand that more is involved than just a question of popular language. We need to understand the political and class content, the alliance of class forces that the slogan points toward.

We are fighting for a government of the workers *and* the farmers, just as the slogan says. We're fighting for a government in which the workers—(exploited wage labor)—and working farmers—(exploited rural commodity producers)—govern together. We are not proposing a government in which workers will govern over our allies, the non-exploiting farmers. This is quite important.

We are proposing a revolutionary replacement of the outmoded bipartisan government of American capital by a government of the parties and leaders of the workers and working farmers that *will* get this country moving again. A government that can organize to produce the industrial goods and food that can improve the lives of people in this country and around the world. That's what we're for. A workers and farmers government.

Being clear on this will help the allies of the working class understand the dictatorship of the proletariat. It's not a dictatorship of the proletariat over the working farmers, over the exploited rural population. It's the end result of a process, of successful governance by the workers and farmers that advances the toilers' interests—*against* the exploiters, *against* the capitalists, *against* the landlords and the banks—to the point where a new economic basis for the government has been established. A government based on expropriation of the exploiters, on nationalized state property, on economic planning—a workers state.

A workers and farmers government organizes the toilers to move toward the expropriation of their common enemies, the capitalists and big landowners. Since both workers and working farmers are exploited by capital, this transitional task directly serves the class interests of both. It's true that wage workers don't share identical class interests with independent commodity producers who own their own tools and the product of their labor. It's true that petty commodity production on the land will produce and reproduce capitalist relations in the rural areas. There will be a class struggle in the countryside in this country too, and the agricultural workers and poorer farmers will be the workers' most stable and reliable ally.

But workers and farmers *do* share a common class interest in sweeping financial, industrial, commercial, and landed capital off the face of this land. And under a government based on proletarian economic foundations, the relations between the countryside and the city, between workers and farmers, between agricultural workers and independent commodity producers, will change over many, many years. Everyone will become associated producers, rather than wage workers, farmers, and so on.

Not a trick

The governmental slogan in our transitional program is not a trick. Communists are not trying to trick the farmers into a workers government by giving it another name. We're fighting for a government of the workers *and* exploited farmers, who together will use that governmental power to transform the economic foundations of society. We say to the farmers, as Trotsky advised us more than 40 years ago, "it will also be your government." And we say so truthfully.

The exact forms that government takes—what kind of committees, councils, and so on—will be decided in practice. But it will be a government organized to do away with exploitation and oppression, as the workers and exploited farmers deepen the class struggle against their common capitalist enemies. It will be the most democratic government ever known in this country.

Ultimately, if the slogan is a trick, then the trick's not on the farmers, but on the working class. Because discounting the importance of an alliance with the farmers is the last thing that the working class can afford to do in making and consolidating the socialist revolution.

Trotsky explained this in the Transitional Program:

> The practical participation of the exploited farmers in the control of different fields of the economy will allow them to decide for themselves whether or not it would be profitable for them to go over to collective working of the land—at what date and on what scale. Industrial workers should consider themselves duty bound to show farmers every cooperation in traveling this road: through the trade unions, factory committees, and, most importantly, through a workers' and farmers' government.
>
> The alliance proposed by the proletariat—not to the "middle classes" in general but to the exploited layers of the urban and rural petty bourgeoisie, against all exploiters, including those of the "middle classes"—can be based not on compulsion but only on free consent, which should be consolidated in a special "contract." This "contract" is the

program of transitional demands voluntarily accepted by both sides (p. 166).

If there is one lesson that we've been taught by the revolutions of our times, it is the importance of this "contract" with the exploited rural producers, made possible through a workers and farmers government.

That is why I disagree with what Comrade Steve Bloom said under the Iran discussion earlier in the plenum. Comrade Bloom defended the decision of the National Committee minority to drop the slogan of a workers and farmers government from its resolution on Iran and replace it with the slogan, "A workers government in alliance with the poor peasantry."

I think the political reasoning behind this change and Comrade Bloom's motivation for it don't apply just to the governmental slogan for Iran—although the seriousness of the error is more glaring, given the size and weight of the peasantry and agrarian question there. I think this relates to the discussion of what our transitional governmental slogan should be in the United States, as well.

Comrade Bloom said during the Iran discussion that the phrase "a workers government in alliance with the poor peasants" in the NC minority resolution is not meant as a popular slogan, but as a "scientific" characterization. By this, Comrade Bloom seems to be saying that in reality the slogan is calling for the dictatorship of the proletariat, to which the poor peasants (working farmers in this country) are "allied." It is not *their* government; it is the proletariat's government. The farmers are merely "allied," "linked to," "supporting."

This is an ultraleft position. It heads toward a repudiation of the transitional use of our governmental slogan. We don't advocate that. We advocate a workers *and* farmers government, and we *mean* it. It is a pledge and a promise to the workers' allies in the countryside. And it is a necessary and powerful step on the road to a socialist America and a socialist world.

Proletarian leadership

Once again, we distinguish between the workers and farmers government and the dictatorship of the proletariat because it is necessary from a *political* standpoint to emphasize the *transition*—the difference between *getting there* and *being there*. That is how a revolutionary proletarian leadership must gauge where it is in the process of consolidating a workers state, how far it has come, how far remains to go, how the class differentiations are developing in city and country, the extent of its progress in holding onto its strategic alliances, and *what tasks* flow from all this. What needs to be done next.

Calling for a workers and farmers government in no way undermines our Marxist understanding of the proletariat's leadership of all the toilers in the socialist revolution and the irreplaceable role of a Leninist working-class party. To the contrary, proletarian leadership is indispensable in the fight for such a government.

That's why our strategic goal is the fight to transform the American labor movement into a revolutionary class-struggle instrument that can point the way forward for the entire working class and all its allies. That's why we say that the American unions must learn to think *socially* and act *politically*. That's why we see the fight for a labor party, for independent working-class political action, as the axis of our transitional program to answer the catastrophe facing the workers and farmers of this country.

That's also why we're determined that the SWP carry out the turn to industry and all its political and organizational consequences. That's why we're determined to proletarianize the SWP as the nucleus of a mass revolutionary workers party that can lead the American socialist revolution. That's why we're determined to defend the extension of the socialist revolution in the Americas, and why we're eager to collaborate with the proletarian leaderships in Cuba, Nicaragua, and Grenada that stand at the head of that revolutionary process.

The line of march of the American working class, in alliance with the farmers and other exploited producers, leads to a revolution to replace the current imperialist government with a workers and farmers government. That is the keystone to the program and strategy of a proletarian party in the United States.

The party constitution

If we adopt the governmental slogan being proposed here by the Political Committee, then we should propose a change in the SWP Constitution, too, at the next convention. Article II under the heading "Purpose" now says:

> The purpose of the party shall be to educate and organize the working class for the abolition of capitalism and the establishment of a workers government to achieve socialism.

When we were discussing with Farrell [Dobbs] in preparation for his testimony at the trial last year, he took a look at that sentence and pointed out that it puts the cart before the horse. Leaving aside for a moment which governmental slogan we use, it should nevertheless be clear that the toilers must first establish a revolutionary government before the abolition of capitalism can be carried out.

So the sentence needs to be turned around. It *should* say that the purpose of the SWP shall be to educate and organize the working class for the establishment of a workers and farmers government leading to the abolition of capitalism and the achievement of socialism. Something like that will be the constitutional amendment we will present to the next convention if we decide here to change our transitional slogan.

Looked at from the standpoint of organizing the workers for the socialist revolution, and recognizing the vanguard role of the working class, the need to change our slogan to "For a workers and farmers government" should be obvious. It was the turn to industry, our new step in the proletarianization of the party, that took us through the experiences that made it possible to come to this decision. It was the working class that won our party to the perspective of this kind of government.

And that's exactly what the American working class will do on a much grander scale in organizing itself and reaching out to its allies with the proposal to join together to fight for a workers and farmers government.

VIII. SUMMARY

As comrades who've studied our material on the workers and farmers government know, I said nothing in the report from the Political Committee that has not been said in writing somewhere before by SWP leaders. I *did,* however, say some things that we have not *adopted* previously. I quoted extensively from Joe's 1978 introduction to Bob Chester's bulletin and from 1970 and 1975 letters from Joe to Bob. These have never been adopted by any body of the SWP.

But, with the agreement of the Political Committee, I intentionally said nothing in the report that we have not said before.

The reason is simple. It's important to pull together the conclusions we have drawn from twenty years of acting, thinking, and writing on the governmental question. This, of course, is a decisive question, because it has to do with the struggle for power.

In the things that we wrote over the twenty years following the Cuban revolution, there are some new things said about the workers and farmers government. And Joe wasn't writing just as an individual, but on the basis of the experiences and thinking in the party leadership on this question.

We do, however, have serious disagreements with the majority of the leaders of the Fourth International on these questions, disagreements on both those questions that have been posed in a new way since Cuba and on those that predate the Cuban revolution. There are numerous different positions among those in the Fourth International who disagree with us, as well. Some comrades hold that there is no such thing as a workers and farmers government; others say that it's a highly improbable theoretical possibility that we've never seen in history; others believe we've seen one in Cuba but nowhere else; still others would add Nicaragua and some would add Grenada. The variations are numerous.

A two-class government?

Comrade Jones, a spokesperson for the United Secretariat Bureau majority, said during the discussion here that there can be no such thing as a two-class government. We disagree. Of course there can be and have been two-class governments. We *advocate* such a government.

A two-class government existed in Russia between October 1917 and the end of 1918, at least. Lenin accurately named it a workers and peasants government, and he wasn't under any illusion that workers and peasants are the same class. He also knew that without the leadership of the working class, led by *its* party, this government would fail.

And the Bolsheviks acted accordingly.

We do *not* believe that there can be two-class parties; a party proletarian in composition, combat experience, and program is decisive in leading the toilers to make the revolution and consolidate its conquests.

We do *not* believe that there can be two-class states; by this we mean that in our epoch states are *either* based on bourgeois or proletarian economic foundations.

But a two-class government? Yes. In fact, it is only in this way that we can understand the dialectics, the transitional character, the use of our governmental slogan. Let's take a closer look at this.

There are not two-class states. The class character of the state is defined by the class character of the existing property relations; that is a fundamental precept of historical materialism.

If the state is based fundamentally on capitalist property relations, as it remains in Nicaragua and Grenada today, then the class character of the state is capitalist *in that sense*. That is *not* the class character of the government, however. (Or of the "state," if that word is being used as a synonym for the government and its repressive forces.)

It is also necessary to remember the other side of the equation, as I explained in my report. We insist that a workers state, the dictatorship of the proletariat, still exists in the Soviet Union today, despite the Stalinist bureaucracy. State property, planning, the monopoly of foreign trade—these economic conquests of the Soviet working class have survived. And they continue to have a revolutionary impact on the course of world history. Confusion on the importance of the process of transition from a workers and farmers government to a workers state can also breed tendencies to downgrade the importance of state property as a historic conquest of the world proletariat, and this can lead to a retreat from our unconditional defense of the workers states against imperialism. I think we can see danger signs of such a tendency in our world movement today, in particular a retreat from our longstanding opposition to the demand for bilateral nuclear disarmament, as opposed to unilateral nuclear disarmament of the imperialist powers.

Yes, we believe that a transitional government is both possible and necessary in the way I explained in my report. Such a government is *not* a workers government in the sense of a workers state, which is what we mean by the dictatorship of the proletariat. It is a government of the workers and exploited farmers that can serve as a bridge to the consolidation of a workers state.

No, there cannot be a two-class government of the *exploited* and the *exploiters*. That kind of government remains a bourgeois government. There can even be bourgeois governments in which a workers party predominates. The Mitterrand government in France today is an imperialist government; it's a bourgeois government presided over by a bourgeois workers party.

But there *can* be a two-class government that represents the interests of the workers and the exploited farmers, as a transition toward the dictatorship of the proletariat, a government that leads the workers toward the expropriation of the bourgeoisie.

Of course, even after the establishment of state property and a planned economy, the government remains a two-class government in an important sense. As the report explained, both wage workers and exploited farmers have a common class interest in expropriating the capitalists. The government that successfully carries out that historic task remains a government of all the producers, under the leadership of the working class. Isn't the revolutionary government of the Cuban workers state a government of the workers and farmers? The solidity of that class alliance remains crucial all the way through the extension of the socialist revolution and the transition to world socialism.

Need for a proletarian party

Does adopting the perspective of a workers and farmers government mean that we should build a two-class party? Comrade Jones implies that it does.

No, to the contrary. Adopting this slogan means we should build a *proletarian* party. A communist party able to unify and lead the revolutionary proletariat along its line of march, including forging an alliance with other exploited producers in the fight to form a workers and farmers government.

The role of leadership by a revolutionary proletarian party is indispensable. Only such a party can advance a program that gives concrete and realistic historical solutions to the problems facing the oppressed and exploited of all classes, along the historical line of march of the proletariat toward becoming the dominant class in society, as opposed to the bourgeoisie, and the eventual abolition of classes.

The less proletarian, the less communist the leadership, the more difficult it is to establish a workers and farmers government and the more likely it is that such a government will be shoved back and defeated, as happened in Algeria. Contrast, for example, the proletarianization of the Castro and Ben Bella leaderships over the two years following the establishment of workers and peasants governments in those countries. Joe pointed out in 1965 that in Cuba "the masses were mobilized by the Castro team which then moved toward the organization of a revolutionary-socialist party *in the very process of the revolution.*" This was not the evolution of the Ben Bella leadership, and that factor was a decisive one in the reversal of the revolution there.

Let me repeat what the 1923 resolution of the Executive Committee of the Comintern had to say on the relationship between the proletarian party and the workers and peasants government, because it remains a good guideline for today.

> Our Party, as far as its composition and aims are concerned, must remain a party of the working class, but of a working class which draws in its wake all sections of the working population and leads them into the fight against capitalism.

Our turn to industry, more than any other single thing, has driven home to us the political centrality of this challenge in building a proletarian party.

Two qualitative turning points

There are *two* qualitative turning points, not just one, in the course of an anticapitalist revolution. The refusal to acknowledge this lesson from socialist revolutions in the twentieth century was one of the things at the root of our difference on Nicaragua with the majority at the 1979 World Congress. We state, definitively, that without the resolution of dual power, there can be no workers and farmers government. That's the *first* qualitative turning point. It was passed in Nicaragua in July 1979, and in Grenada in March of that same year. It was resolved in Cuba by mid-1959.

We also state that there must be a *second* qualitative turning point. That is, a workers and farmers government, regardless of its leadership, cannot expropriate the bourgeoisie without mighty mobilizations of the toiling masses themselves. When the new economic relations have become the dominant ones through that process, a workers state has been consolidated.

Parenthetically, I should add that we disagree with Comrade Jones's statement that there is no such thing as a peasant party. Maybe this is just a terminological problem, I'm not sure.

But, yes, there can be a peasant party, and there have been and are today. There can be a party that is primarily peasant in composition and that has a petty-bourgeois leadership claiming to speak for the peasantry. Of course, in carrying out this claim to speak for the peasantry, such a leadership will in reality present a perspective either along the lines of the solutions of the bourgeoisie or along the lines of the solutions of the proletariat. The peasantry can have no independent class program.

But saying this does not mean that such a party is simply a bourgeois party or simply a proletarian party, merely claiming to speak for the peasantry. If it is a party that is overwhelmingly peasant in composition, it will have important particular attributes. Different parties of such a type will have programs reflecting demands of different layers of the peasants. An alert proletarian leadership will at times incorporate and use certain of these demands, as the Bolsheviks did in the October revolution.

Trotsky wrote in the 1930s that the Chinese Communist Party had "actually [torn] itself away from its class" and became a party with a mass peasant base and petty-bourgeois leadership. No longer a proletarian party, it became a peasant party in this sense. It was that kind of party that led the revolution to victory in 1949. Of course, the Chinese working class welcomed the triumph over the hated capitalist government of Chiang Kai-shek. Many workers joined the CCP, despite the new regime's repressive acts against and constraints on workers' struggles. Moreover, the workers and peasants regime had to rely on mobilizations of the workers—in order to combat counterrevolution and the threat of imperialist invasion in the early 1950s and carry out the transition to a workers state. A process of proletarianization was necessary in China, too, although the CCP remained the party of a petty-bourgeois caste and the Chinese workers state was deformed from birth.

The Khmer Rouge under Pol Pot is another example of a peasant party, one whose political trajectory, as we know, was quite different from what occurred in China.

Of course, if you mean by a peasant party one that can offer *a different class solution* to the crisis of our epoch, then the answer is "no." Any such party will either look toward the capitalists and their program, or to the proletariat and its program.

But there *can* be a two-class government, a government of the proletariat and the exploited rural producers. If it is led forward by a proletarian leadership—or even by a petty-bourgeois leadership that ends up mobilizing the workers under extreme circumstances, as in China in the early 1950s—it can move forward to the expropriation of the capitalists.

Once again, there can't be a state based on two sets of property relations. It is either based on capitalist property relations or it rests on state property as a result of the expropriation of the capitalists. A workers and farmers government still rests on the property relations it inherited while it seeks to organize and mobilize the toilers to transform those relations. That's why we believe Nicaragua remains a workers and farmers government, one still on the road of organizing, mobilizing, educating, and inspiring the urban and rural toilers to move forward to the expropriation of the bourgeoisie.

Relationship of class forces
I want to deal with the five questions raised by Comrade Frank Lovell during the discussion.

His first question was: How viable is a workers and peasants government? How long can it last before capitalist property relations have been overturned as the dominant form of economic relations?

That depends upon many concrete factors: the strength of imperialism and its capacity to intervene militarily on the side of the counterrevolution; the political caliber of the leadership; the relationship of class forces domestically and internationally; and so on. These will vary widely from one situation to another. There's no blueprint or timetable.

The leadership must know how to size up the situation. It must be able to deepen the popular mobilizations in such a way that the class differentiations and crises that will inevitably occur will be resolved in a progressive, not a reactionary manner.

It was the Ben Bella leadership's incapacity to pursue a consistent revolutionary course along these lines that led to the defeat in Algeria. It failed by backing off from the mobilizations in the countryside, refusing to lead the unions forward, and failing to establish a popular militia. That political vacilation and retreat set the stage for the bourgeoisie and nascent bourgeoisie to use the army officer corps under Boumedienne to step in and overthrow the Ben Bella government in 1965. By that time, the workers and farmers government had been almost completely eroded by the lack of a leadership capable of moving it forward.

So the viability depends upon factors such as these, which boil down to whether the relationship of class forces is shifting toward or away from the workers, both at home and abroad. I repeat, the second qualitative leap—the one necessary to establish a workers state—can't just be *administered* from above by the workers and farmers government. The workers and their allies must be educated, armed, mobilized, and organized. It is a process.

We followed this revolutionary process closely in Cuba, giving the precise time and circumstances under which a workers and farmers government came into existence in mid-1959 and explaining how that government organized and led the masses to deepen the revolution until a workers state had been established by autumn 1960. We also followed the process in Algeria that eventually culminated in the stillbirth of that revolution.

And since the revolutions in Grenada and Nicaragua in 1979, no other current in the workers movement has devoted as much political attention, time, and effort as we have in charting the course of those workers and farmers governments and their proletarian leaderships, so that we can learn the lessons that only living revolutions can teach.

Of course, once a workers state has been consolidated, the danger of counterrevolution does not disappear. A workers state can still be overthrown. The workers will struggle to defend state property and their other conquests, but only the victory of the world socialist revolution can provide a final guarantee against successful capitalist counterrevolution.

Lessons from the Bolsheviks

A second question that Comrade Lovell raised was: What is the relationship of this workers and farmers government to the Bolsheviks' goal of a revolutionary democratic dictatorship of the proletariat and peasantry?

Well, there are two sides to answering that question.

On the one hand, the Bolsheviks' formula is a model of the transitional method that Lenin used to build a proletarian party capable of leading the Russian workers and peasants to power in October 1917. Like the workers and farmers government slogan, it correctly captured the strategic class alliance that the workers had to forge in order to lead the revolution to victory. It gave the correct answer to the class content of the victorious revolutionary government. In that sense, the two slogans have similar ties in their fundamental strategic concept and use.

On the other hand, Lenin described the revolutionary democratic dictatorship of the proletariat and peasantry as the formula used by the Bolsheviks as "a Marxist definition of the class content of a victorious revolution" *in Russia*. He never claimed it to be a governmental slogan for the world, and he rarely used it after the October revolution.

The slogan of the workers and farmers government as first discussed and adopted in the Comintern, however, is a different matter. That *was* projected by the Bolsheviks at the Fourth Comintern Congress and the subsequent ECCI meeting as a governmental slogan with worldwide application. They saw it as a summation of what they had accomplished in the Russian revolution, the kind of government they had set up, its relationship to the expropriation of the bourgeoisie, the strategic alliance with the peasantry, and revolutionary developments from Germany to Bulgaria to Czechoslovakia to other places around the world. And this was not just a slogan for backward countries, nor just for imperialist powers such as Russia, still burdened by remnants of feudalism, medievalism, and landlordism. It was an international perspective.

The workers and peasants government was a transitional concept for building a bridge to the dictatorship of the proletariat, or a bridge from the opening stage of the dictatorship of the proletariat through the expropriation of the bourgeoisie, in a world political situation where Communists still had before them the task of winning a majority of the workers and other toilers to a revolutionary perspective and a revolutionary party in *every* country. Defeats had already been sustained in Germany and Hungary by the time of the Third Comintern Congress, and in Italy as well by the time of the Fourth Congress. So this was a key aspect of the united-front policy and transitional approach adopted by the Comintern as part of its efforts to orient revolutionary communists toward winning the urban and rural masses away from social democratic and other petty-bourgeois and bourgeois misleaders.

Does it go beyond Trotsky?

A third question raised by Comrade Lovell: Does what's being proposed here go beyond Trotsky?

Yes. It's certainly true that Joe went beyond Trotsky in his introduction to Bob Chester's bulletin referred to in the report. He went beyond Trotsky because he had to deal with some thirteen revolutions that led to the establishment of workers states after Trotsky died. And quite a few revolutions that didn't. The SWP and the Fourth International went beyond Trotsky, too, when we analyzed the social overturns in Eastern Europe, China, and Cuba. So we've all gone beyond Trotsky.

That's not Trotsky's fault. And it's not our fault. In fact, in that same introduction to Bob Chester's bulletin, Joe explained what Trotsky said and didn't say on this question.

"Trotsky referred to [the workers and farmers question] in passing in the Transitional Program," Joe wrote, "but he did not enlarge upon it."

But we did. We *had* to, for reasons Joe went on to explain.

The necessity to resume where the Fourth Congress of the Communist International had left off arose from new complex events in the international class struggle. In the aftermath of World War II, workers states appeared in Eastern Europe, China, Korea, Vietnam, and Cuba. The processes through which these states came into being had to be explained correctly in the light of Marxist theory.

Trotsky's passing reference served us well in beginning to grapple with this question. It gave us a clue where to look and what to look for, so that we could once again pick up the thread of the workers and farmers government and really *put it to use*. This is what we began to do following the Cuban victory in 1959.

The theoretical problems posed for us hadn't been solved by the 1952 IEC report on China, which incorrectly sought the key to the workers and peasants government in "elements of dual power which still remain in China." Nor were these problems addressed by the resolution on China adopted at our September 1955 NC plenum and our 1957 convention, where we recognized that a badly deformed workers state had been established, but offered no theory of the transitional government.

No, it was going back to the Comintern discussions and to the clues left by Trotsky in the light of the Cuban revolution that subsequently led us to review our approach to the Chinese and Yugoslav revolutions. That's what we did in the report presented by Joe for the Political Committee and adopted at our 1969 convention.

Bob Chester had taken part in the discussions on China in the 1950s, and he was clearly struck by the extent of the change marked by the 1969 report from our previous reports and resolutions. So Bob dropped Joe a letter, pointing out that the report "adds something new to our theory of the transformation of bourgeois states into workers states," and that he "would have taken exception" to these changes had he been able to attend the convention.

In his letter replying to Bob, Joe said he thought it was true that, "our resolutions did not go into the details of the process involving the transfer of power" from a capitalist state to a workers state. The one exception was Cuba, Joe said.

Bob was convinced by Joe's answers to the questions he had raised, and he began to look more closely himself into these revolutions; one of the results was Bob's study on *Workers and Farmers Governments Since the Second World War,* which he had almost completed at the time of his death in 1975.

So, the Cuban revolution was the turning point for us in beginning to understand the workers and farmers government and how revolutionists could advocate it as a powerful tool of the working class in leading a socialist revolution and carrying it through to the consolidation of a workers state. This opened the door for us to apply what we had learned from Cuba to the Algerian revolution a few years later, as well as to the Chinese and Yugoslav revolutions, where we had left some loose ends.

It should be no mystery why Cuba was easier for us to understand. It had a revolutionary leadership, so it didn't seem like such a difficult problem.

To understand why it was so difficult before that, you've got to put yourself back three decades, to the time when one-fifth of humanity was ripped out of the capitalist world by a revolution *under Stalinist leadership.* Then you can begin to understand the challenge to the continuity of our theory, and our need to work through the questions posed by it.

But Trotsky had left us a clue in the Transitional Program, as I said. He wrote:

> . . .one cannot categorically deny in advance the theoretical possibility that, under the influence of completely exceptional circumstances (war, defeat, financial crash, mass revolutionary pressure, etc.), the petty-bourgeois parties, including the Stalinists, may go further than they themselves wish along the road to a break with the bourgeoisie.

We correctly never tried to use this sentence from Trotsky as some open-and-shut mandate for the conclusion we came to on Cuba. It didn't flow directly at all. Because Trotsky had immediately followed up that sentence with this one:

> . . . even if this highly improbable variant somewhere, at some time, becomes a reality and the workers' and farmers' government

in the above-mentioned sense is established in fact, it would represent merely a short episode on the road to the dictatorship of the proletariat (p. 175).

Trotsky's assumption—and the assumption of the Fourth Comintern Congress discussions, as well—was always that without a Bolshevik-type party winning the leadership, a workers and farmers government could not go all the way to the establishment of a workers state. That was our view until the end of the 1940s.

But that's not how history turned out. A number of things happened during the decade after Trotsky died that he couldn't have foreseen.

The postwar social overturns in Eastern Europe gave us some problems initially, but they were not the most difficult of these events for our movement to understand, because of the direct role of the army of the Soviet workers state. There, Trotsky had left some more substantial clues in *In Defense of Marxism,* writing on the Soviet occupation of Poland and Finland in 1939. Yugoslavia was more difficult, because there the Soviet army was not directly involved. Nonetheless, the Soviet victory over German imperialism was a decisive factor in the victory of this revolution, too.

The *big* challenge to our theory, as I said, came with the 1949 victory of the Chinese revolution under the leadership of the Stalinist CP. That was unanticipated. We couldn't pretend to read that outcome into Trotsky's single sentence from the Transitional Program. We didn't try to read the outcome in Cuba into it, either.

Instead, learning what we could from Trotsky's analysis and from that of the Comintern, we had to develop our own analysis of these postwar revolutions. And we believe that the views we have developed, and are taking a step further in systematizing in this report, are in continuity with the Comintern positions that Trotsky based himself on in the Transitional Program.

Permanent revolution

The fourth question by Comrade Lovell was: How does this relate to the permanent revolution?

Well, this is what we think the permanent revolution is. It is the strategic view of the world socialist revolution that integrates the proletarian revolution in the economically advanced capitalist countries, the democratic, anti-imperialist and socialist revolution in the colonial and semicolonial countries, and the struggle to defend, extend, and democratize the workers states. It is the proletariat's fight, under the leadership of a Leninist party, to unify and organize the toilers in wresting governmental power from the exploiters in a massive revolutionary upsurge and then using that governmental power to deepen the class struggle and go forward to their expropriation. No other class can lead even the bourgeois revolution through to the end.

That's the perspective for our time. That's the communist perspective. That's the perspective that the Socialist Workers Party and the Fourth International have been fighting for since our founding.

The victory of the world's first workers state, and the inability of imperialism and domestic counterrevolution to crush it during three years of civil war, marked a qualitative shift in the international relationship of class forces. No aspect of the world class struggle was left untouched by it. The October revolution transformed the prospects for successful anticapitalist revolutions everywhere.

Prior to 1917, this had been ruled out for countries with overwhelmingly peasant populations such as China, India, and Afghanistan. With help from Soviet Russia, however, the Comintern now saw the possibility on a world scale of workers and peasants governments coming to power and serving as a bridge to the expropriation of the bourgeoisie.

That's exactly what we're talking about here. What we call soviet power is *not* the dictatorship of the proletariat, a workers state, right off the bat. That would only be true if the soviets were entirely proletarian and if the expropriation of the capitalists were immediate. But both of those conditions are excluded in every country of the world.

In Lenin's 1920 speech to the Second Comintern Congress on the colonial and national question, he stressed the possibility and importance of peasant soviets, as well as workers Soviets.

> The idea of Soviet organization is a simple one, and is applicable, not only to proletarian, but also to peasant feudal and semi-feudal relations.

> Our experience in this respect is not as yet very considerable. However, the debate in the commission [on the colonial and national question], in which several representatives from colonial countries participated, demonstrated convincingly that the Communist International's theses should point out that peasants' Soviets, Soviets of the exploited, are a weapon which can be employed, not only in capitalist countries but also in countries with pre-capitalist relations, and that it is the absolute duty of Communist parties, everywhere to conduct propaganda in favour of peasants' Soviets or of working people's Soviets, this to include backward and colonial countries. Wherever conditions permit, they should at once make attempts to set up Soviets of the working people.

The perspective of peasants soviets, Lenin continued, raises a closely related question:

> The question was posed as follows: are we to consider as correct the assertion that the capitalist stage of economic development is inevitable for backward nations now on the road to emancipation and among whom a certain advance towards progress is to be seen since the war? We replied in the negative. If the victorious revolutionary proletariat conducts systematic propaganda among them, and the Soviet governments come to their aid with all the means at their disposal—in that event it will be mistaken to assume that the backward peoples must inevitably go through the capitalist stage of development.
>
> Not only should we create independent contingents of fighters and party organisations in the colonies and the backward countries, not only at once launch propaganda for the organisation of peasants' Soviets and strive to adapt them to the pre-capitalist conditions, but the Communist International should advance the proposition, with the appropriate theoretical grounding, that with the aid of the proletariat of the advanced countries, backward countries can go over to the Soviet system and, through certain stages of development, to communism, without having to pass through the capitalist stage.
>
> The necessary means for this cannot be indicated in advance. These will be prompted by practical experience. It has, however, been definitely established that the idea of the Soviets is understood by the mass of the working people in even the most remote nations, that the Soviets should be adapted to the conditions of a pre-capitalist social system, and that the Communist parties should immediately begin to work in this direction in all parts of the world ("Report of the Commission on the National and the Colonial Questions, July 26, 1920," *Collected Works*, Vol. 31, p. 240–5).

So, the *October revolution* opened up a new possibility for the entire world, including for the colonies and semicolonies. The revolutionary toilers will set up soviets of the peasants and other working people, Lenin said. They'll try to do it like the Russians! Our workers and peasants republic will aid them.

And that created the possibility of a new kind of government on an international scale.

But what we are proposing for consideration here today is not something that can be found in its entirety in the Comintern documents or in the Transitional Program. Joe's sentence that "the first form of government that can be expected to appear as the result of a successful anticapitalist revolution" is a workers and farmers government—that is *not* in the Transitional Program. It couldn't have been, given the historical experiences of the workers movement up to that time.

But that *is* the conclusion we draw from the postwar period, especially from the twenty years following the Cuban revolution. That's the conclusion Joe drew looking back on Cuba, and that we can affirm and generalize watching the course of the revolutions now unfolding in Nicaragua and Grenada. These two revolutions put our theory to the test and verified it. It confirmed the views we had been developing since 1959, based on our continuity with the Bolshevik-led Comintern.

A necessary stage?

Is a workers and farmers government a necessary stage in the development of an anticapitalist revolution? That was Comrade Lovell's fifth question.

Our answer, based on the history of revolutions since 1917, has to be "yes." That's what we say—it's the "first form of government we can expect." And such a government is a necessary stage in *this* country, too, which is the world's largest, most powerful, and wealthiest capitalist power.

Do we think we can have an instantaneous workers state in this country? Instantaneous expropriation of the bourgeoisie and implementation of all-encompassing economic planning? Stop to think for a moment what governing this country is going to be like in the period after an anticapitalist revolution. Think about what putting this country back together will be like after the capitalist class has waged a civil war trying to crush a workers and farmers government.

Given the high level of industrial and overall economic and social development in the United States, the workers and farmers here will certainly face much more favorable conditions to carry out the transition to new, proletarian economic foundations than in any successful revolution to date. But the workers and farmers government will still confront the task, under whatever concrete conditions it inherits, of organizing, mobilizing, and educating the working people to expropriate the bourgeoisie, institute workers control, expand workers management as the basis for economic planning, and govern the country. In the process, the class struggle will deepen and differentiations will take place, culminating in the consolidation of a workers state that will initiate the transition to socialism.

If the workers and other urban toilers want to be fed during this transition, they will need a sound alliance with farmworkers and exploited farmers in this country. Moreover, given the productive potential of U.S. agriculture, American farmers will contribute mightily to the construction of a rational and humane society on a world scale.

I repeat, there are no two-class parties. All "two-class" parties are fakes; those that claim to be are actually parties that serve the interests of the bourgeoisie.

We're building a proletarian party. Of course, we don't restrict its recruits only to workers. We want exploited farmers to join the proletarian party. We want anyone who comes over to the side of the workers' struggle for socialism to join our ranks.

But it's a proletarian party that must be built to lead the American socialist revolution. A party proletarian in membership, leadership, combat experience, and program. A party with a large component of Blacks and Latinos, immigrant workers, and women in its membership and leadership. That's what we're out to build.

And there are no two-class states. In fact, it is only by recognizing this difference between the class character of governments (defined by their program, composition, and the class interests they advance) and the class character of states (defined by their economic foundations) that enables us to understand why we should advocate a workers and farmers government as the indispensable bridge from a capitalist to a workers state.

Our 1967 change

Finally, there are a couple of other points I want to deal with.

One comrade during the discussion expressed the opinion that we had made a "foolish" change in our governmental slogan in 1967. I don't think that is accurate. That's an ahistorical way of looking at the decision. The report explained that the impetus behind the change was our attempt to relate politically to the rise of the Black struggle in this country in the 1960s. It was part of the proletarianization of the SWP, which has been a constant effort by the party. That was the impetus, and it showed we were a serious proletarian party, alert to developments in the struggles of our class and its allies.

But in 1967 we simply weren't far enough along in the development of the U.S. class struggle and the proletarianization of the party to arrive at the decision that we propose to make today. We were just beginning to shake out a few other effects of our semisectarian existence during the postwar retreat of our class. So we couldn't totally get on top of the farm, rural, and agriculture questions. We couldn't do that until the turn.

But by correctly explaining the transitional character of our governmental slogan, and by rejecting alternative slogans that would have led us more off course, the motion adopted at our 1967 convention prepared us to correct our mistakes given our further experience over the subsequent fifteen years.

This is important. Because it's always easy to denigrate the road we've traveled if we look back at our past ahistorically. Think about how somebody could look back on *us*, at *this* National Committee, in fifteen years. We will certainly have more experience and know a lot more in 1997 than we do now. Looking back with fifteen years under our belt, we'll even know more about 1982. But this National Committee will deserve some credit for the party's achievements over those fifteen years. And the same goes for looking back to 1967, too.

We need to be able to evaluate our own past experience as materialists, in order to move forward on a sound political footing. This is how, by applying our program and strategy in life, we simultaneously build on and thus enrich our revolutionary continuity.

The American Theses

Let me give another example. At the beginning of the plenum, the Presiding Committee distributed a statement dated February 27 from Comrades Henderson and Weinstein. I don't know whether this is their new platform, or an addition to the platform they submitted to the 1981 convention. Either way, my point here is only to quote one paragraph from it about the "Theses on the American Revolution" adopted at our 1946 convention.

I've also noticed that Comrades Lovell and Bloom have also been raising the American Theses more and more, as if invoking the authority of this important party document somehow settles any of the disputed political questions facing the party today.

"The basic ideas in [the American Theses] retain their full force and provide an accurate guide for the future of socialism in the United States," Comrades Henderson and Weinstein write in their statement.

I disagree with that. It's a one-sided statement. The American Theses don't provide an accurate guide for the future of socialism in the United States today on at least one key strategic and programmatic aspect of the American revolution. It doesn't include the combined character of the American revolution.

There are altogether only four sentences in the American Theses that deal with the Black question. Four sentences. That paragraph correctly explains that Blacks increasingly have come into industry and into the CIO, AFL, and other unions, and that they are part of the vanguard of workers' struggles in this country (James P. Cannon, *The Struggle for Socialism in the "American Century"* [New York: Pathfinder Press, 1977], p. 343 [2012 printing]). That's all there is.

The Chicano, Puerto Rican, and other Spanish-speaking population is not mentioned at all. The "Americanization" of a proletariat that had formerly been language-divided was correctly noted by the resolution in 1946, but thirty-six years later, the U.S. working class is again composed of a substantial percentage of non-English-speaking workers, especially those who speak Spanish. Nor does the resolution, given when it was written, deal with the transformations that have occurred over the last few decades with regard to the role of women in the working class.

The resolution was also written in a period in which we could not anticipate the failure of the post–World War II revolutionary upsurges in Western Europe and how that would significantly slow down the American revolution. But Farrell [Dobbs] explained at the party's very next plenum in 1947 that we had actually adopted the American Theses in the midst of what turned out to be the start of a decline of the class struggle in this country; it turned out that the American revolution *had* been postponed—for thirty-five years, as of now.

None of this detracts from the American Theses as a key part of our revolutionary continuity, above all on one decisive question for the American proletarian party. It rejected and rebutted any idea of American exceptionalism, that is, any idea that the American revolutionary perspective had to be indefinitely suspended because of the economic and military power of American capitalism and the relatively higher living standards of the American proletariat. It affirmed the American socialist revolution as a realistic perspective for our time; that is its great and most lasting accomplishment. And it affirmed that the SWP is the irreplaceable nucleus of a mass proletarian party that can lead the American working class and its allies to power.

But the American Theses alone is *not* an accurate guide for the future of socialism in the United States. To say that is also a totally ahistorical statement, just as it would be ahistorical to reject the American Theses as part of the continuity of our movement because of these shortcomings measured against what our class faces today.

Go back and reread the American Theses and see if you agree with Comrades Henderson and Weinstein that they "provide an accurate guide for the future of socialism in the United States." I think you'll see that they don't. And then go back and look at some of our convention resolutions from the past ten or fifteen years, where we correctly explain that these more recent resolutions couldn't have been written without the fundamental perspectives I've pointed to from the American Theses.

That's how a workers party learns from and builds on its experiences to preserve and develop its revolutionary continuity. That's what we're doing by adopting this report and deciding to adopt the slogan "For a workers and farmers government" as a central part of our program and strategy for the American socialist revolution.

Our decision here today will also be a contribution to the discussion in the Fourth International. This question not only involves our differences with others in our world movement over the Nicaraguan, Grenadian, and Cuban revolutions and their leaderships, but also our differences over the turn to industry and the related political and organizational questions of party building; the fight against imperialist war; and our defense of the workers states and state property. It goes to the root of the transitional program and strategy necessary to build proletarian parties and a proletarian international.

The Political Committee is placing two motions before the plenum:

1) That we change our transitional slogan, "For a workers government" to "For a workers and farmers government"; and

2) to approve the general line of the Political Committee report and summary.

[The vote on the first motion was: regular: 39 for, 2 against, 0 abstentions, 0 not voting; consultative: 39 for, 0 against, 0 abstentions, 0 not voting.]

[The vote on the second motion was: regular: 38 for, 2 against, 1 abstention, 0 not voting; consultative: 38 for, 0 against, 1 abstention, 0 not voting.]

Appendix I

Introductory note to 'Workers and Farmers Governments Since the Second World War'
by Joseph Hansen

The following is the introduction, written in 1978, by Joseph Hansen to *Workers and Farmers Governments Since the Second World War,* an Education for Socialists bulletin, by Robert Chester.

This study deals with a subject that to many socialist militants might appear at first sight as hardly of great concern: *What is the first form of government that can be expected to appear as the result of a successful anticapitalist revolution, and how does it relate to the preceding struggle for power?*

The topic itself came under consideration quite late in the development of key revolutionary-socialist concepts. It was submitted for general discussion for the first time at the Fourth Congress of the Communist International in 1922. Only the delegates of the Bolshevik Party, in the period when it was led by Lenin and Trotsky, could have suggested the importance of the question to the cadres of the Third International.

The delegates at the Fourth Congress did not engage in fanciful speculation. Their debate was based on the experience of the October 1917 revolution in Russia, on five years of thinking over that mighty chapter in the development of civilization, and on the need to bring subsequent experiences into the context of the lessons of 1917.

After 1922 the subject was not taken up again. The life-and-death struggle with Stalinism cut across further development of Marxist theory on this question as on much else. Trotsky referred to it in passing in the Transitional Program, which was adopted at the founding congress of the Fourth International in 1938, but he did not enlarge upon it.

The necessity to resume where the Fourth Congress of the Communist International had left off arose from new complex events in the international class struggle. In the aftermath of World War II, workers states appeared in Eastern Europe, China, Korea, Vietnam, and Cuba. The processes through which these states came into being had to be explained correctly in the light of Marxist theory.

Failure to do so would have put in question the continuity of Marxist theory, including Trotsky's analysis of the meaning of the extension of the borders of the Soviet Union at the beginning of World War II and eventually his analysis of the degeneration of the first workers state.

To carry out this task, the significance of the post–World War II overturns of capitalism had to be connected with the conclusions reached by the Fourth Congress in 1922. Those conclusions had to be either rejected, extended, or modified as the facts might dictate.

The importance of the question becomes obvious when it is thought through and the consequences for political practice are grasped. Nonetheless, it is a fact that it remains a field of prime interest only to advanced revolutionary cadres. This holds true for the world Trotskyist movement as a whole.

The main reason for this discounting of the question is to be found, I think, in the paramountcy of problems facing small revolutionary organizations in disseminating a revolutionary-socialist outlook among the masses. A better understanding of what is involved can be gained if we single out three general aspects, or phases, of this consciousness-raising process—not forgetting, of course, that in the final analysis they mesh together:

1. The educational work of bringing the masses to understand that the great social and economic evils they suffer from are consequences of capitalism in its death agony, and that the dilemma facing humanity on a world scale with ever-increasing acuteness is *socialism or barbarism*. The task is preeminent in countries where the program of revolutionary socialism is represented by only small minority movements.

2. The organizational work of building a revolutionary-socialist mass party as the means

for meeting the central dilemma. The problem facing small revolutionary groups of linking up with the masses comes under this heading. The task demands doggedness, the utmost attention, and an expenditure of time and effort bordering on fanaticism.

3. The final push of playing a leading role in the working-class struggle for power when the conditions for this have matured.

For periods longer than expected, revolutionists have had to concentrate on the two preliminary phases. The associated tasks are just as difficult as those of the third phase—perhaps more so. The preliminary problems, standing in some instances for years, if not decades, at the top of the revolutionary agenda, can certainly appear to be more real than the question of what form of government might appear as the consequence of a revolutionary victory.

However, in today's highly unstable world, seemingly remote theoretical questions have a way of suddenly imposing themselves in the political arena and demanding answers that can decisively determine the fate of groups and currents bidding for leadership of the working class. Thus problems related to the struggle for power cannot be placed in deep-freeze to be brought out "when the time comes." They are with us now, both in the sense of internationally important events on which stands must be taken (the Cuban victory, for instance), and in the sense of gaining a more concrete appreciation of the possibilities in coming struggles.

Moreover, the struggle for power, along with the accompanying problems and tasks, must be kept constantly in mind. As the goal, that culminating phase dominates our decisions in selecting the means required for its realization.

Bob Chester was one of the cadres of the Socialist Workers Party who saw the importance of studying that feature of a socialist revolution called a "workers and farmers government." He set out to gain an independent understanding of the phenomenon, going back to the Russian experience and moving to subsequent events in other countries.

He had not finished his study when he died of a heart attack on June 22, 1975. The manuscript he left was thus somewhat rough. Perhaps in a final draft he would have dealt with some points at greater length while compressing others or placing them in a different order.

Certainly he would not have changed his views. The more material he gathered and thought over the more convinced he became of the importance of the topic. I am sure he would have felt deep satisfaction if his study succeeded in helping others to gain the insights he achieved through this work even though they might not agree with everything he said or the way he put it.

Appendix II

Letter from Joseph Hansen to Robert Chester

The following 1975 letter from Joseph Hansen to Robert Chester comments on an initial draft of Chester's study, *Workers and Farmers Governments Since the Second World War*. The page numbers and numerous formulations cited by Hansen refer to the manuscript of this draft, not to the version published as an Education for Socialists bulletin following Chester's death several months later in 1975.

New York, N.Y.
April 18, 1975

BOB CHESTER
SAN FRANCISCO

Dear Bob,

I am sorry about the delay in getting to your manuscript "Workers and Farmers Governments Since the Second World War." With events breaking as fast as they have recently, it's difficult to turn to subjects that require more leisurely consideration. In any case, I finally got to it.

First of all, I am impressed by the amount of work you put into the project. You must have devoted more than your spare time to it. The material you present from the period of the Russian revolution is especially useful. And I think you must have gone over the draft a number of times, since it is quite condensed, yet covers most of the angles quite well. I really don't have a great deal to say, other than a few general observations plus a few small points:

1. Until I came to the very last section "Conclusions," I kept waiting for a succinct description of the category "Workers and Farmers Government" in relation to the capitalist economic relations it stands on, the destruction of capitalism, and institution of planned economy to which it can eventually shift as its new base. You have a few sentences in the "Conclusion"; but it might be good to explain this fully enough in the first sections so as to make it quite clear before you get into the examples. Mentioning it in the "Conclusion" could serve as a summary.

It would be still better, in my opinion, if you would, besides this, repeat the explanation in each of the examples you cite—but in concrete terms. Cuba offers the best example, inasmuch as every stage stood out with crystal clarity. The facts spoke for themselves, so to say.

China is the most difficult one, and I am not sure that you have completely solved it. For instance, your description of the Yenan regime as an "agrarian democracy" is not clear to me. Precisely what does this mean? Especially the term "democracy"—under *Mao*?

Considering the size of the population involved, more than ninety million (much larger than Brazil at the time), the Yenan economy and regime offer a topic of considerable complexity and importance that has been insufficiently studied from the Marxist point of view.

Related to this is the question of the Asian-type economy, which Marx separated out as a distinct category. Where did this fit in? As a system eroded by the penetration of capitalism from the West and from Japan? Did it play no role in the territories governed by Yenan and the assembling peasant armies?

Such questions come up, it appears to me, in tracing the appearance and evolution of the Chinese workers and farmers government.

Also I note that you leave out the role of the army, particularly its administrative function. But the striking parallel between the role of this army and previous peasant armies in the ancient cycle of Chinese revolutions really ought to be dealt with.

Again, in your handling of China, it might be useful to bring out more clearly (by indicating in more detail) the role of the Korean war in building the pressure for finishing with capitalism and introducing a workers state.

2. Do you have some reason for leaving out the cases of Vietnam and Korea? In the case of Eastern Europe you indicate the role of the occupying Red Army and explain that the overturns there deserve special study. So why do you leave out Vietnam and Korea?

3. In the case of Cuba, it would be helpful to supply the dates of the process. I haven't checked to make sure, but I seem to remember having put them in the record in debating Tim Wohlforth or Bert Deck (maybe both of them).

Now for some small stuff, noted in passing:

Page 1, line 2 from bottom: "necessary to full-fledged workers states." A degenerated workers state, then, from a full-fledged condition would have lost its feathers. It might be better to say "demanded by the Leninist norms for workers states." Also I note in this sentence that you leave out the item of toppling capitalism (which is required before they can emerge as workers states).

Page 41, lines 12 to 14 from top: The formulations appear one-sided to me. A military victory would not have been possible in China in the absence of a profound social upheaval. The armies were the cutting edge of the blade, but only the edge. The city workers welcomed the PLA (not just "passively lined the streets") even though the CCP made no effort to mobilize them and they were not in the leadership.

Page 49, lines 17 to 19: Here, as I believe elsewhere, the characterization is so condensed as to suggest that a workers and farmers government evolves *into* a workers state or represents the first phase of a workers state. For clear analysis the categories should be separated, the better to indicate the role that can be played by a workers and farmers government—beginning on the basis of the capitalist economy and even part of the capitalist state structure, destroying and replacing the capitalist state structure, establishing a monopoly of foreign trade, expropriating the key industries, introducing a planned economy, etc., and finally ending up as a regime (good or bad) standing on a workers state.

Page 69, last paragraph: Another of the sentences that is too condensed. ". . . economic transition from a workers and farmers government to workers state and workers government . . ." would be easier to follow if you develop it a bit, particularly if you bring in the step regarding what happens to the capitalist economy and capitalist state.

Page 71, line 10 from top: "speed." Ought to add "clarity." Speed and clarity. That is clarity up to the point where Moscow's influence became heavy.

Page 75, line 5 from top: Leave out "classical." What is a "classical guerrilla situation"?

Page 83, line 9 from bottom: "Raptis [Pablo] . . . had become an adviser to Ben Bella. . . ." I think this is inaccurate. Raptis sought political asylum in Algeria. (He had real need for it when he was released from prison in Holland. Had he returned to Greece, he would certainly have been imprisoned by the dictatorship.) His reputation was very high in FLN circles because of his role as a leader of the Fourth International in organizing material aid in Europe for the Algerian revolution. When Pablo arrived in Algiers, Ben Bella made it a point to see him. Asked how he expected to make a living, Pablo said he would like to get a teaching job in the university. Ben Bella said he would see what could be done. Later Ben Bella asked Raptis to come and see him. Ben Bella asked if instead of teaching he would be willing to work in the *biens vacants* administration as they were desperate for help there. Pablo agreed. Later he was asked to work in the offices handling the agrarian reform.

Pablo did his best to get others to join him in Algiers. He hammered particularly hard on Ernest Mandel, saying it was a crime for him to stay in Brussels when a person of his abilities was needed so much in Algiers, not only because of the needs of the Algerian revolution, but specifically because of the needs of the Fourth International.

Many of Pablo's enemies have pictured him as an "adviser" to Ben Bella, particularly Healey; thus the story got around. The most I ever heard along these lines from his most devout disciples was that in one or more speeches on agrarian reform Ben Bella relied heavily on memorandums drawn up by Pablo. In the department where he was working, Pablo may have had some influence among those occupied there. He had more substantial influence with Harbi and others engaged in putting out the paper (its name was changed several times).

I might add that Pablo flew to Chile during the Allende regime. He again got around into various government departments. But no one thought to call him one of Allende's advisers. I suppose they didn't know he was there.

Do you really need to credit Pablo in this way? After all, he was a representative of the Fourth International at this time and it was the Fourth International that had the reputation. Why reduce it to Pablo and then convert it into an innuendo?

Page 85, line 5 from bottom: "autogestion." The English is "self-management."

Page 88, line 3 from top: "direction." Course?

Page 89, line 4 from bottom: "advisers." Wouldn't it be better to characterize them as members of the left-wing of the FLN? They represented one of many political currents that Ben Bella tried to bend with, neutralize, or straddle, etc. The picture of a sultan with viziers is misleading.

Page 92, line 12 from bottom: "autogestion." Use the English.

Page 94, line 11 from bottom: A typo. 1964 not 1944. (In general the copy editing is excellent. I caught a few obvious mistakes in spelling.)

Page 94, line 7 from bottom: "adviser." Do you need to mention Raptis here? Others were more prominent—and more influential. A *current* was being isolated and defeated.

Page 94, line 5 from bottom: Why "maquis" rather than guerrillas?

Page 103, lines 7 to 11 from top: In speaking about the role of the workers in Cuba, shouldn't you indicate the special weight of the *sugar* workers?

Page 111, lines 9 to 15 from top: This makes the problem of a date, which is very important from the viewpoint of dialectics, seem altogether too vague. It seems to me it would be better to say that the dates on which qualitative changes occurred can be placed within quite narrow brackets. We did this, as you know, in the case of Cuba. The quote from Trotsky deals with the question of *continuity* and not the points of qualitative change involving discreet categories, which can be pegged or closely bracketed as to date.

Page 111, last sentence in paragraph 3: "Mao never attempted to analyze why the same policies as those he followed in China before the victory, failed so tragically in Indonesia."

As formulated, this is dubious. Mao, of course, never analyzed. But where was the Indonesian Yenan, the formation of peasant armies, the resistance to imperialist invasion, the civil war? Mao OK'd the class-collaborationist policies of Aidit in the *Indonesian* situation. Both Aidit and Mao were class collaborationists in the pattern going back to the first betrayers in the modern class struggle. So turn your point around, why didn't Aidit's policies, which failed so tragically in Indonesia, fail just as tragically in China? Well, they did in 1925–27. . . .

In light of this, the previous sentence about the Yugoslavs ought to be restudied.

The following sentence, too, is open to misinterpretation. "None of the leaderships . . . could measure up. . . ." Is one to conclude that while they didn't measure up, still they were in the same league—no doubt down towards the bottom but still in the same league?

Page 113, line 4 from top: "Their adaptation to the Soviet Union hardened all the deformations of their revolution." Hardened deformations . . . a caste in Cuba? Perhaps this should be discussed; but it is really a separate subject from the one you are dealing with, and needs more than a declarative sentence. Perhaps it would be better to leave it out, holding it for discussion elsewhere.

That's about all.

With our best
Joe

P.S. I am turning the manuscript over to the Education Department. Maybe they will have additional suggestions.

Appendix III

An exchange of letters between Joseph Hansen and Robert Chester

The following exchange of letters between Joseph Hansen and Robert Chester took place following the 1969 convention of the Socialist Workers Party. The exchange was first published in *The Workers and Farmers Government,* an Education for Socialists bulletin, by Joseph Hansen.

NOTE ON HANSEN-CHESTER CORRESPONDENCE

by Bob Chester

This correspondence began as a partial misunderstanding of Joe Hansen's report to the 1969 convention (*Internal Information Bulletin* No. 4 in 1969), a misunderstanding that was cleared up in the course of the interchange. Joe viewed the formation of a workers and peasants government with the victory of Mao as a special case, and not as part of a pattern that would have designated workers and peasants governments in Yugoslavia and the buffer zone with the end of World War II, and in Cuba in January 1959.

There were a number of theoretical questions dealing with this topic that I felt needed clarification, and I utilized this opportunity to raise them. While the interchange has been in hiatus for the last few years, both Joe and I feel that the discussion is still continuing.

There is no doubt that the topic is important for the theoretical arsenal of our movement. It is a fruitful area for study and research, to which many comrades can make contributions.

AUGUST 26, 1973

LETTER FROM BOB CHESTER TO JOSEPH HANSEN

December 13, 1969

Dear Joe,

I want to raise a question on your report to the convention, "The Origin of the Differences on China," printed in Bulletin No. 4. While I consider the report as a whole an excellent one and would have supported it if I were at the convention, I am sure that I would have taken exception to the conclusions you presented on workers and peasants governments. [The relevant portions of this report are reprinted in this collection.]

As far as I know this is the first time we have claimed that Yugoslavia and the buffer countries had workers and peasants governments beginning with the defeat of the Nazis in 1945; that China had a workers and peasants government with the Mao victory in 1949; and that Cuba had a workers and peasants government with the victory of Jan. 1, 1959. I do not remember it being included in our resolutions covering these events and I believe it adds something new to our theory of the transformation of bourgeois states into workers states in colonial and semicolonial countries.

Previously we had concluded that what had been set up in the East European countries were coalition governments of native Stalinists, direct agents of Moscow, together with peasant, social democratic, and whatever capitalist elements they could find. The Red Army controlled and guided their operation. The Soviet Union plundered these countries of many basic factories, let capitalism operate on a national level, even set up joint stock companies with them, and paid little attention to the needs and wishes of the workers. These governments played a repressive rather than progressive role. Could these be considered as workers and peasants governments independent of the bourgeoisie? I think not. Even when the transition to a workers state took place as a reaction to the Marshall Plan, it was done on a controlled basis under the direct aegis of the Soviet bureaucracy. We termed this process "structural assimilation," and the result deformed workers states. Would you characterize the governments that effected these changes workers and peasants governments? If you do you would have to qualify them even further to account for the fact that they were acting as agents of the Soviet bureaucracy.

In the country having the most advanced mass participation in the revolutionary process, Yugoslavia, I do not think the Tito-Subasic government could be classed as a workers and peasants government. Even though it was short lived it was essentially a coalition government of Stalinists and capitalists set up under the pressure of the imperialists and Moscow. It is true that the government did not last long, that the bourgeois elements found little room to operate, and that with the failure of the coalition the Tito regime swung sharply left. From that point on the designation of workers and peasants government might apply.

In Cuba, the initial government set up with Urrutia as president could only be characterized as a coalition government, even though its program gave promise of more radical change. The institution of agrarian reform brought this government into crisis, ending with the ousting of Urrutia in the fall of 1959. The Draft Theses of Dec. 23, 1960 under point 7 begins with the statement, "The fact that Cuba *now* had a Workers and Farmers government. . . ." (my emphasis) reinforces the thinking that it became such a government with the expulsion of the capitalist wing.

In China, as your report details, Mao came to power at the head of a peasant army and took over the cities with the conscious policy of preventing the working class from playing any significant role in the revolution. They brought native capitalists into the government and announced a policy of peaceful and friendly relations with capitalist governments, including the United States. Land reform, which had been a feature of the pre-victory period, was curtailed and the Peoples Democracy they visualized was one that would continue for a prolonged period. It was the Korean events with the threat of U.S. invasion that spurred the leftward shift that ended in the formation of the deformed workers state. You raise the question of how we would designate the Mao regime in the years before the 1949 victory and pose the possibility of a workers and peasants government in the years before that victory. If that were so it seems to me that a shift in the manner of rule occurred with the national government set up in 1949.

It is true that our concept of the type of government that existed in the transition period has never been fully developed. There has always been a period during the leftward shift, before we could designate these states as workers states (deformed or otherwise) with workers governments at their head, that we used the term "Workers and Peasants Governments." Even then we never clearly established what the distinctive characteristics of these governments were nor did we identify them as the necessary "link in the revolutionary process" that established these workers states. It seems to me that this is precisely the problem we have to solve.

When a revolution takes place in a colonial or semicolonial country with the active participation of the masses in support of a leadership that can be either worker, peasant or middle class, can we immediately designate this as a workers and peasants government? What are the characteristics that make it "worker and peasant government" rather than radical bourgeois or peasant? What would be the points of qualitative change of one to the other? Above all, what are the dynamics of a worker and peasant regime that make it the "link in the revolutionary process?" I know that these questions have not been answered to my satisfaction.

Neither can I see how the theses and discussions at the Fourth Congress of the C.I. are really applicable here. They deal essentially with the tactics used by a revolutionary party during a period of crisis when it does not yet have hegemony over the working class. The purpose of these tactics, which are basically that of critical support and the united front, are to force the reformist governments leftward, expose the reformist leaders and speed the process of radicalization of the working class and its allies. The objective was the formation of a workers and peasants government under the leadership of the revolutionary party that would in essence be the dictatorship of the proletariat.

It is true that the Congress listed four different types of workers or workers and peasants governments, and listed them under these classifications in order to set their tactical orientation to each. It also pointed to the fact that "liberal workers governments" such as the British Labor Party and "social democratic workers governments" as in Germany were in essence "coalition governments of the bourgeoisie and anti-revolutionary labor leaders" who would have to be overthrown in the course of the struggle.

In the light of the Theses of the Fourth Congress how could you possibly explain a petty bourgeois leadership that emerges out of the national struggle and *as a body* with comparatively few defections goes through the transition to a workers state? I see this as a verification of the theory of permanent revolution and not of the Theses of the Fourth Congress. If they can be classed as workers and peasants governments, then they are new and special types which require independent analysis.

I have tried to think this problem through and, after generalizing the whole experience on the question, have come up with some different answers. I would like to get your reaction to them.

It seems to me that the solution lies in a closer examination of the national liberation struggles under the special set of circumstances that have existed from the end of World War II. They include the considerable weakening of imperialism on a world scale; the enhanced strength and prestige of the Soviet Union; the great weight and power of the colonial struggles for liberation; the reduced size, strength and loss of confidence of the native capitalists. These were bolstered in former periods by the weight of the compradore bourgeoisie, but these sections now flee at the early stage of struggle leaving the native capitalists even weaker. Add to this the chronic crisis of the peasantry.

A revolutionary movement rising out of the national struggle, even under middle class leadership, sees imperialism as its main enemy and also sees a counterbalance to it—the attractive power of the Soviet Union and the other workers states. The movement feels that imperialism no longer has such overwhelming power that it cannot be opposed or defied. There is now room for maneuver. It is no accident that there has emerged a "third world" section, the so-called neutrals, that play both sides in an attempt to gain some advantages for themselves.

The national character of the movement gives the leadership an added advantage, in that it can gain support of workers and peasants as well as the middle class and some section of the capitalists in its fight against imperialism. If there was a conscious revolutionary party on the scene this "national unity" would not exist but would separate into sharply contending wings of the class struggle. The process would then take a different road, following more closely the pattern of struggle laid out by Trotsky, where the capitalists and its reformist supporters would be progressively isolated and the workers with their peasant allies would move toward power, to a workers and peasants government or directly toward a dictatorship of the proletariat. With the absence or failure of a revolutionary party to rise to the situation a petty bourgeois leadership can get the support of diverse elements and come to power on a limited, short term program meeting the new situations in its typical empirical fashion.

The attitude of the imperialists is crucial. If it is hostile and adamantly opposes the leftward swing, especially nationalizations of "its" holdings, the revolution has the option of turning to the Soviet Union for aid. Where the leadership had Stalinist origins as in Yugoslavia, China, and Vietnam, this reaction would be on the order of an automatic reflex. In Cuba the turn to the Soviet Union went hand in hand with the sharpening of the struggle with Washington. If the imperialists adopt a more flexible attitude and make concessions in order to maintain some economic ties the process can be slowed down or even reversed. I think Algeria is an example of this. The lever France used was that of economic agreements, especially on the exploitation of Sahara oil.

Nationalization is a logical step for a backward country that must find the means of setting up enterprises for national survival or for trade on the world market. There is no other national source to finance them. Taking over the property of the huge imperialist monopolies is a natural starting point for national independence. The recent examples in Bolivia and Chile indicate how strong this pressure is even on bourgeois leaders friendly to imperialism. Where the masses participate in the revolutionary process nationalization is far more sweeping, especially where Soviet aid and advice become large factors in the nationalization process. Control of foreign trade and banking become natural logical steps as measures of defense if not as positive policies to defend the gains already achieved.

Thus the struggle, beginning at the level of national liberation, with the support of the masses moves on past the national level into the permanent revolution. What is essential for the process to develop into the workers state form is the maintenance of the pressures that started the revolution in the first place; a national revolutionary upsurge, a leadership responsive in some measure to the mass pressure, hostility to imperialism concretized in nationalizations, elimination of capitalist elements that oppose this process and the continued pressure of imperialism that prevents any stabilization and adaptation to a peaceful coexistence at some intermediate stage. With this also develops a socialist consciousness which has either been in the background of the movement or has been acquired in the course of the struggle.

Does a government that carries out these measures have to be a workers and peasants government? Could it not be carried out by a petty bourgeois government or even a national coalition of anti-imperialist forces, at least in the early stages? The experiences of the twenties and thirties indicated that when the middle class was caught between class pressures the majority invariably gravitated toward capitalism. With the relationships changing in the post–World War II period it is more possible for sections of the middle class, when under the pressure of the mass movement, to gravitate toward the working class. This has apparently happened in a few special cases.

It is clear that the negative factors persist in this development. While they follow the main line of permanent revolution the transition is not thorough nor complete. They result in deformed or incompleted workers states and remain on the national level. Where they go beyond the national level, it is solely as a measure of national defense. While the leaderships have opposed or eliminated threats from the right they equally oppose tendencies from the left that try to move the struggle to the international level.

This is the main outline of my thinking on the problems posed by your report to the convention. I admit that I have not thought through a number of theoretical and tactical problems raised by them. That is why I am posing them on a tentative basis. I believe they are important enough to be included in the general discussion that is now underway.

I am not raising these ideas as a necessary projection for the future, only as a generalization of what has happened. It is quite possible that we are entering a new stage that will not follow the previous pattern. The French events of 1968 are one sign of a change. There is also a new awareness by U.S. imperialism of its failure in the cases of China and Cuba to prevent the formation of hostile workers states, that might express itself in a shift in approach to new revolutionary movements. There is also a growth of revolutionary Marxist trends in some of the colonial countries that could sharply change the character of the national struggles. Any one of these factors can have a strong effect upon the type of revolutionary forces that emerge on the scene of colonial struggle.

I hope to get your views on my criticism of your convention report as well as on the other views I have raised.

Comradely,
Bob Chester

LETTER FROM JOSEPH HANSEN TO BOB CHESTER

July 26, 1970

Dear Bob,

First of all I want to express my appreciation for your patience in waiting such a long time for me to reply to your letter of last December 13. As I explained when I had a chance to talk with you about this, I wanted to go back and check particular items you had raised. I especially had in mind going in some detail into several of the overturns of capitalism that have occurred in circumstances other than under the leadership of a revolutionary-Marxist party. Unfortunately I have not been able to make time for this although it is certainly required if an adequate response is to be written to the many important questions you have raised. Finally I decided that further delay was simply impermissible and that in the absence of the research that ought to be done, I could write down my reactions to your letter. It would be easiest if I could just take your letter paragraph by paragraph. This is rather awkward, but how else can we proceed like a conversation? So get out your letter to refer to as I go along. To help out some, I will number the items and quote extensively from what you say.

1. "As far as I know this is the first time we have claimed that Yugoslavia and the buffer countries had workers and peasants governments beginning with the defeat of the Nazis in 1945; that China had a workers and peasants government with the Mao

victory in 1949; and that Cuba had a workers and peasants government with the victory of January 1, 1959. I do not remember it being included in our resolutions covering these events and I believe it adds something new to our theory of the transformation of bourgeois states into workers states in colonial and semicolonial countries."

In response to this I think it is true that our resolutions in general dealt only with the main question of the toppling of the capitalist state in those countries and its replacement by a workers state of one kind or another (generally "deformed"). Our resolutions did not go into the details of the process involving the transfer of power. As to the dates you indicate, I will not go into these here because, as I said, I have not been able to make time to dig into details. In the case of Cuba, however, I will make an exception. In the Winter 1961 issue of the *International Socialist Review* I wrote:

> On coming to power, the July 26 movement set up a coalition government that included well-known bourgeois democratic figures—and not in secondary posts. In retrospect these may have seemed middle-class decorations or mere camouflage hiding the real nature of the government. It is more accurate, I think, to view this government as corresponding to the political aims of the revolution as they were conceived at that time by its leaders.
>
> But such a government stood in contradiction to the demands of the insurgent masses and to the commitment of the July 26 movement to satisfy these demands. The Revolution urgently required far-reaching inroads on private property, including imperialist holdings. As Castro and his collaborators moved toward fulfillment of the agrarian reform they met with resistance from their partners in the coalition, a resistance that was considerably stiffened by support from Wall Street which viewed them as "reasonable" elements in a regime packed with bearded "wild men."
>
> As Huberman and Sweezy correctly observe, "a sort of dual system government began to emerge." The displacement of Felipe Pazos by Che Guevara in November 1959 marked a decisive shift and the resolution of the governmental crisis, whatever hang-overs from the coalition still remained. The government that now existed was qualitatively different from the coalition regime.

I then went on to develop the view that this was a "Workers and Farmers Government" and that this government initiated and carried out the measures that brought the workers state into being in Cuba "between August–October, 1960. . . ."

To continue with your letter:

2. "Previously we had concluded that what had been set up in the East European countries were coalition governments of native Stalinists, direct agents of Moscow, together with peasant, social democrats, and whatever capitalist elements they could find."

As we have seen in the case of Cuba, the setting up of a *coalition government* does not exhaust the process. The question under examination requires an analysis in finer detail. We can understand what is involved if we ask, does the appearance of a *coalition* government open the possibility of a socialist overturn in the absence of a revolutionary-Marxist party? If so, what will most likely be the process by which such an overturn will occur?

I agree with your observations about the real power in Eastern Europe being exercized by the Red Army and that this was the decisive force in the overturns. However, I think you overlook an important link in the organization of local committees and indigenous governments. Some of the figures in these governments proved not to be "direct agents of Moscow." This was shown shortly thereafter by the extensive purges in Eastern Europe and the jailing and execution of various prominent leaders. I seem to recall Trotsky warning of such a possibility when he was discussing the Finnish events at the beginning of World War II, or perhaps it was the Polish events. In the light of the uprisings in later years in East Germany, Poland, Hungary, and the struggle in Czechoslovakia, it would be worth reexamining in detail the process by which capitalism was toppled in Eastern Europe to fix in a more precise way the role of the native Stalinists in distinction from the direct agents of Moscow. However, as I had to admit earlier, I couldn't make time for this.

3. "We termed this process 'structural assimila-

tion,' and the result deformed workers states."

The term "structural assimilation" was always somewhat puzzling to me.

When I was in Europe in 1962 I took up this question with the comrades who had first used the term. They said they meant something simpler than we had imagined in translating what they had written. "Structural assimilation" to them merely referred to the fact that the economic and state forms that had appeared in Eastern Europe were modeled on those of the Soviet Union. "Structural assimilation" thus occurred also in the case of Yugoslavia and China, likewise in Cuba, although to a lesser degree.

But the comrades in Europe did not follow the process in the finer detail that concerns us in the question of the role of workers and peasants governments under the conditions we have specified.

4. "Would you characterize the governments that effected these changes workers and peasants governments?"

The problem is one of substance. Did indigenous governments exist in these countries? If so, what role did they play in the transfer of power and the establishment of deformed workers states? Their role may not have been much; but evidently it was sufficient to lead Stalin to decapitate them—and at a rather early stage.

5. "If you do you would have to qualify them even further to account for the fact that they were acting as agents of the Soviet bureaucracy."

I would agree to this if you in turn would agree that while being agents they were also at the very same time not agents, or at least had the potentiality of not being agents. The case of Tito is outstanding but the same holds for others like Rajk whom Stalin succeeded in liquidating.

6. "In the country having the most advanced mass participation in the revolutionary process, Yugoslavia, I do not think the Tito-Subasic government could be classed as a workers and peasants government. Even though it was short-lived it was essentially a coalition government of Stalinists and capitalists set up under the pressure of the imperialists and Moscow. It is true that the government did not last long, that the bourgeois elements found little room to operate, and that with the failure of the coalition the Tito regime swung sharply left. From that point on the designation of workers and peasants government might apply."

This is one of the items in your letter that I would have liked to check out in order to determine more specifically the points of qualitative change. Among the things of special interest: (a) Which wing was dominant in the coalition? (b) When the qualitative change occurred from a coalition government to a government so radical in nature that it was capable of destroying the capitalist state structure (and the capitalist economic forms) and establishing a workers state, then it would seem obvious that not only was the date important but also the nature of the new government.

This is precisely the key point under discussion. You say: "From that point on the designation of workers and peasants government might apply." With this sentence haven't you in essence conceded the case I have been arguing for? For if it "might apply" to Yugoslavia, why might it not also apply to the other cases? Once you have made such a concession in a *single* instance, it appears to me insuperably difficult from a methodological point of view to maintain that it be excluded in all *other* instances where comparable governments have similarly ended capitalism and established workers states.

But then why the hesitancy? Why "might apply" instead of "does apply"?

7. "In Cuba, the initial government set up with Urrutia as president could only be characterized as a coalition government, even though its program gave promise of more radical change. The institution of agrarian reform brought this government into crisis, ending with the ousting of Urrutia in the fall of 1959. The Draft Theses of Dec. 23, 1960 under point 7 begins with the statement: 'The fact that Cuba *now* had a Workers and Farmers government. . . .' (my emphasis) reinforces the thinking that it became such a government with the expulsion of the capitalist wing."

I, of course, agree with your observations outside of the item as to the point of qualitative change and I would not argue much about that. The ouster of Urrutia could well be considered to have marked the point of qualitative change. At the time this occurred, however, I preferred to wait to see what the positive consequences of the negation might be. Caution was dictated, I thought, because of the absence of any declared anticapitalist and prosocialist program of the Castro wing. When Che Guevara was moved into a key position and then took charge of carrying through the changes that brought the first workers state into being in the Western hemisphere, that appeared to me to mark an unquestionable point of qualitative change as determined on the level of *action*.

8. "In China, as your report details, Mao came to power at the head of a peasant army and took over the cities with the conscious policy of preventing the working class from playing any significant role in the revolution. They brought native capitalists into the government and announced a policy of peaceful and friendly relations with capitalist governments, including the United States. Land reform, which had been a feature of the pre-victory period was curtailed and the Peoples Democracy they visualized was one that would continue for a prolonged period. It was the Korean events with the threat of U.S. invasion that spurred the leftward shift that ended in the formation of the deformed workers state. You raise the questions of how we would designate the Mao regime in the years before the 1949 victory and pose the possibility of a workers and peasants government in the years before that victory. If that were so it seems to me that a shift in the manner of rule occurred with the national government set up in 1949."

I do not disagree with your general description. The question of a "workers and peasants government" falls *within* this general description. For instance, you say: "It was the Korean events with the threat of U.S. invasion that spurred the leftward shift that ended in the formation of the deformed workers state." The question I raise is "leftward shift" of *what*?

To answer this question adequately, it appears to me necessary to make a detailed study of the facts, considering them in their actual historical sequence, and going back to the period before the 1949 victory in view of the prior existence of the Mao team and its rule over a considerable territory for a considerable period of time. To me it appears obvious that the nature of the Mao government in the pre-1949 days is involved.

Likewise involved is the question you point to; that is, the "manner of rule" of the national government set up in 1949. This ought to be examined within the context of the previous manner of Maoist rule. I do not doubt that a "shift" occurred—the extension of Maoist rule over all of China is an obvious instance as is the expropriation of "bureaucratic" capital. It appears to me that what would be most valuable for our movement is a detailed study of this question. I sought to suggest only a few general guidelines for this study which might in the end have to be greatly modified or junked in the light of the concrete facts.

9. "It is true that our concept of the type of government that existed in the transition period has never been fully developed. There has always been a period during the leftward shift, before we could designate these states as workers states (deformed or otherwise) with workers governments at their head, that we used the term 'Workers and Peasants Government.' Even then we never clearly established what the distinctive characteristics of these governments were, nor did we identify them as the necessary 'link in the revolutionary process' that established these workers states. It seems to me that this is precisely the problem that we have to solve."

I agree with this almost wholeheartedly. I say "almost" because my memory is hazy on how we designated those governments in Eastern Europe. I wanted to check back to see if we said anything. My impression is that we did not really attempt to state the problem or solve it until the Cuban revolution occurred. I may be mistaken in this; perhaps you have already gathered the material necessary to verify this.

10. "When a revolution takes place in a colonial or semicolonial country with the active participation of the masses in support of a leadership that can be either worker, peasant, or middle class can we immediately designate this as a workers and peasants government?"

I would say, "No." I have in mind specifically the case of Nasserite Egypt.

11. "What are the characteristics that make it a 'worker and peasant government' rather than radical bourgeois or peasant?"

I would say that the chief characteristic is its direction of movement. This is indicated by its words (declared program) and its actions. The actions are decisive and we should discount the words if they prove not to coincide with the actions of the government. Some notable examples are available for study in this respect—Nasser's socialist demagogy, for instance, in contrast to his use of state power to foster a new capitalist class; and in the case of the Cubans the opposite contrast, assurances in the first stage about maintaining property relations while their actions were to the contrary.

12. "What would be the points of qualitative change of one to the other?"

I think that we have to regard a "workers and farmers government" in the sense we have been using it as a highly transitional phenomenon. The establishment of such a government by no means leads inevitably to the establishment of a workers state as we have seen in the case of Algeria.

What is most decisive is its practice in relation to the capitalist state structure on which it rests. If a government calling itself "socialist," as in the case of Nasser's regime, simply restaffs the old state structure and intervenes in the economic structure along the lines of "statism," its direction of movement is clearly not toward establishment of a workers state. The social context is also of key importance—the involvement of the masses on a revolutionary scale is required, for this is what basically determines the direction of movement.

The relations with the imperialist powers are also fairly indicative. In the case of Algeria, for instance, the role played by imperialism in overthrowing Ben Bella and in bolstering Boumedienne was very revealing. The captains of world capitalism are exquisitely sensitive on such questions.

As to the actual points of qualitative change, we have already discussed these above in several instances, particularly Cuba.

13. "Above all, what are the dynamics of a worker and peasant regime that make it the 'link in the revolutionary process'?"

What is involved is governmental power. A party or team that gains governmental power thereby gains the possibility of smashing the old state structure and overturning capitalism.

If a revolutionary Marxist party exists, and gains governmental power under the impulsion of a revolution, there is no question as to the subsequent dynamics. The party assures it through its program, through the cadres imbued with that program, and through the experience gained in the living class struggle that finally puts it in power. The course of the Russian revolution is a classic example. Note well, however, that the Bolsheviks held power for a period on the basis of the capitalist state structure and the capitalist economy. Time was required to carry out their program. If anything, they had to carry through these changes *prematurely*. (This had to be paid for later, as Trotsky explained, by the New Economic Policy.) Thus the Russian revolution provided the world with the first example of a "Workers and Peasants Government" in power with the task still before it of actually establishing a workers state.

14. "Neither can I see how the Theses and discussions at the Fourth Congress of the C.I. are really applicable here. They deal essentially with the tactics used by a revolutionary party during a period of crisis when it does not yet have hegemony over the working class."

I agree that this was the main question discussed at the Fourth Congress and that the main outcome was the application of the united front policy for the purposes you indicate.

15. "It is true that the Congress listed four different types of workers or workers and peasants government, and listed them under these classifications in order to set their tactical orientation to each."

This was where it appeared to me that the discussion at the Fourth Congress did have a connection with the problem that faced us. I also added Trotsky's comments in the Transitional Program, which you do not mention.

Under point No. 13 above, concerning the "dynamics of worker and peasant regime," I list the case of such a government controlled by a revolutionary Marxist party. There is no problem for us in this instance; but the question that faced us was the appearance of similar governments in which a party comparable to the Bolshevik party either did not exist or existed as a minority. (In China it existed as a tiny minority that was brutally liquidated by the Maoists.) The Bolsheviks in their discussion excluded the possibility of such governments, controlled by petit bourgeois parties, actually establishing workers states. They reached this conclusion largely on the basis of their experience, an experience, of course, that was determined by the development of the international class struggle and the balance of world power as it stood in their day.

We were confronted, however, by the cases of Yugoslavia, the other countries of Eastern Europe, China, North Vietnam, North Korea, and finally Cuba. We were faced with actual situations never encountered by the Bolsheviks although they had anticipated the possibility (especially Trotsky) in a very general and abstract way. (Yet to me this anticipation was highly illuminating and I once again felt what giants our predecessors were and how open they were to what life itself might bring up, and how prepared they were to make adjustments in the general forecasts advanced on the basis of previous theory and experience.)

Thus we really had no choice, if we were to live up to the norms established by our teachers, except to work out our own answer to the appearance of a decisive "link in the revolutionary process"—the link of governmental power—in which a revolutionary Marxist party was not in control, yet which led to the establishment of a workers state (deformed or otherwise).

The problem was very important in my opinion. If we did not succeed, we were faced with the following alternatives:

Either (a) the Stalinist parties proved to be genuine revolutionary parties after all, even under Stalin. The consequence of this would be inescapable—Trotsky was wrong. This was the line of reasoning followed by Arne Swabeck in the case of China.

Or (b) petit bourgeois forces of no matter what kind and under no matter what kind of circumstances can conceivably establish a workers state and even along the "cold road." This was the line of reasoning of those who came to the conclusion that Egypt, Syria, etc., are workers states, or, under a different label ("state capitalism," etc.) equivalents of the Soviet Union.

Or (c) the facts are inexplicable theoretically and the whole thing must be regarded as a hopeless mess. Perhaps this is the line of thinking of some of the Healyites to judge from the distaste they display in dealing with this problem.

It occurred to me that the discussions at the Fourth Congress of the Communist International, plus the intimation in Trotsky's point in the Transitional Program (under the heading "Workers and Farmers Government") about "the petit bourgeois parties, including the Stalinists," going further "than they themselves wish" along this road although it was a "highly improbable variant," offered valuable clues to a solution of the problem on the theoretical level.

It also occurred to me that if we were to decide that the "highly improbable variant" mentioned by Trotsky had actually occurred, then the events which we were compelled to explain in any case if we wished to remain true to scientific socialism might themselves offer fresh insights and an enrichment of the bare abstractions posed as excluded as a highly improbable variant by the Bolsheviks and Trotsky. I centered my efforts on the Cuban revolution which appeared to me to offer the clearest and most telling example. The results, I think, were not unfruitful.

What was especially instructive was to see the political *differentiations* that occurred in this peculiar variant and the *limitations* on the dynamics of this specific link as it actually developed in the revolutionary process. I don't want to go into that here—I have already discussed it in articles on the Cuban revolution but I mention it as a preliminary to your next question.

16. "In the light of the Theses of the Fourth Congress how could you possibly explain a petit bourgeois leadership that emerges out of the national struggle and *as a body* with comparatively few defections goes through the transition to a workers state?"

As a body. But that is not what has occurred. In the case of Cuba, which offers the clearest example, the July 26 Movement split wide open. An entire wing—the right wing—disintegrated. Some very prominent leaders during the revolutionary struggle against Batista broke from Castro and Guevara to finally end up with the gusanos. The left wing, on the other hand, moved toward socialism and finally declared the Cuban revolution to be a socialist revolution.

To be noted with special attention: This differentiation occurred *after* the July 26 Movement had come to power and while it was dealing with governmental problems on the basis of a still existing capitalist state.

17. "I see this as a verification of the theory of permanent revolution and not of the Theses of the Fourth Congress."

I agree with the first part of your sentence. What happened in Cuba under the workers and peasants government headed by Fidel Castro certainly does verify the theory of permanent revolution.

I disagree with the second part of your sentence. In my opinion, the existence of this peculiar transitional government in Cuba offered the most striking proof of the prescience of the leaders at the Fourth Congress in foreseeing such a possibility (even though they excluded that it could actually establish a workers state). It offers just as striking proof of Trotsky's prescience in the Transitional Program when he left it open as a "highly improbable variant."

18. "It seems to me that the solution lies in a closer examination of the national liberation struggles under the special set of circumstances that have existed from the end of World War II. They include the considerable weakening of imperialism on a world scale; the enhanced strength and prestige of the Soviet Union; the great weight and power of the colonial struggles for liberation; the reduced size, strength, and loss of confidence of the native capitalists. These were bolstered in former periods by the weight of the compradore bourgeoisie, but these sections now flee at the early stage of struggle leaving the native capitalists even weaker. Add to this the chronic crisis of the peasantry."

In this paragraph you do not deal with the specific problem but with the general context in which the problem is located. What you say about this context is true. However, I would place more stress on the general weakening of world capitalism, on the impasse of imperialism, and I would add as a primary part of the context the *default* of Stalinism. The default of Stalinism is required to explain why the liberation struggles have taken the form of *national* liberation struggles instead of *socialist* liberation struggles in such a prominent way in recent decades.

19. "A revolutionary movement rising out of the national struggle, even under middle class leadership, sees imperialism as its main enemy and also sees a counter-balance to it—the attractive power of the Soviet Union and the other workers states. The movement feels that imperialism no longer has such overwhelming power, that it cannot be opposed or defied. There is now room for maneuver. It is no accident that there has emerged a 'third world' section, the so-called neutrals, that play both sides in an attempt to gain some advantage for themselves."

You are still dealing with the context. The paragraph appears rather abstract to me. What examples do you have in mind? Would you include Egypt? What you say would seem to hold true not only for revolutionary movements arising out of the national struggle, "even under middle-class leadership," but also for the bourgeoisie. India, for example.

20. "The national character of the movement gives the leadership an added advantage, in that it can gain support of workers and peasants as well as the middle class and some section of the capitalists in its fight against imperialism."

Again, you are dealing with the general context of the problem and not the problem itself. What you say holds also for the bourgeoisie. It is sufficient to cite the expropriation of the oil industry in Mexico under Cárdenas in the late thirties. Cárdenas was very popular with the workers and peasants both in Mexico and throughout Latin America. A current example is the Velasco regime in Peru with its expropriations, agrarian reform, and anti-imperialist stance.

21. "If there was a conscious revolutionary party on the scene this 'national unity' would not exist but would separate into sharply contending wings of the class struggle. The process would then take a different road, following more closely the pattern of struggle laid out by Trotsky, where the capitalists and their reformist supporters would be progressively isolated and the workers with their peasant allies would move toward power to a workers and peasants government or directly toward a dictatorship of the proletariat."

It is hard to disagree with what you say. If this had been the situation, the revolution would have taken a quite different course in Yugoslavia, Eastern Europe, China, North Vietnam, North Korea, and Cuba. And we would have had no tough theoretical problem to bedevil us.

22. "With the absence or failure of a revolutionary party to rise to the situation a petit bourgeois leadership can get the support of diverse elements and come to power on a limited, short term program meeting the new situations in its typical empirical fashion."

This is precisely the situation that requires theoretical explanation. What is a "petit bourgeois leadership" that finds itself "in power" except a "workers and farmers government" in the sense that we have been using it? If you really hold to this position, what do you object to, then, the nomenclature? But I called attention to this unfortunate nomenclature in the article mentioned above, published in the Winter 1961 *International Socialist Review*. On the other hand, if you really hold to this position, aren't you in contradiction with the substance of your paragraph (No. 1 above) stating that you do not recall such a position having been included in our resolutions and that you believe it adds something new to our theory?

I would not insist on the contradiction but it would seem that if you really hold to this position then you are logically compelled to relate it to the previous theory held by our movement; or, if there has been no previous theory, to explain the reason for its absence. I am convinced that if you were to follow through on this (leaving aside the question of nomenclature) that you would have to go back at least to the Fourth Congress of the Communist International where the question *did* come up.

23. "The attitude of the imperialists is crucial. If it is hostile and adamantly opposes the leftward swing, especially nationalizations of 'its' holdings, the revolution has the option of turning to the Soviet Union for aid. Where the leadership had Stalinist origins as in Yugoslavia, China, and Vietnam this reaction would be on the order of an automatic reflex. In Cuba the turn to the Soviet Union went hand in hand with the sharpening of the struggle with Washington. If the imperialists adopt a more flexible attitude and make concessions in order to maintain some economic ties the process can be slowed down or even reversed. I think Algeria is an example of this. The lever France used was that of economic agreements, especially on the exploitation of Sahara oil."

Again, you are describing features of the context of the problem. I would call attention only to the sentence about its being an "automatic reflex" for leaders of Stalinist origins to turn to the Soviet Union. There are new elements in the context— the rebellions in East Germany, Poland, Hungary, Czechoslovakia, not to mention the Sino-Soviet rift. We should add Cuba's experience, recalling what Guevara said in Algiers about the obligation of the strong "socialist" powers to help the weaker ones and to give aid to revolutionary movements abroad. It is true in general, however, that the leaders of any upsurge will turn in the direction of the USSR, or China, or Cuba, etc., in search of material aid against imperialism. They will turn in any direction that material aid can be obtained. Nkrumah, let it be recalled, helped some of the African movements. So has Boumedienne.

24. "Nationalization is a logical step for a backward country that must find the means of setting up enterprises for national survival or for trade on the world market. There is no other national source to finance them. Taking over the property of the huge imperialist monopolies is a natural starting point for national independence. The recent examples in Bolivia and Chile indicate how strong this pressure is even on bourgeois leaders friendly to imperialism. Where the masses participate in the revolutionary process nationalization is far more sweeping, especially when Soviet aid and advice become large factors in the nationalization process. Control of foreign trade and banking become natural logical steps as measures of defense if not as positive policies to defend the gains already achieved."

We remain within the general context. How to explain the fact that not every government, whatever the pressures, responds so logically to these logically necessary steps? Only workers and peasants governments have thus far proved capable of proceeding in logical enough fashion to end up with the establishment of workers states. Why do other governments, faced with the same pressures, take the course of increasing "statism"?

I do not disagree with what you say although I would not be so categorical in stating that there is no other national source to finance enterprises or trade on the world market except through nationalization. The peasants (leaving aside the workers) remain a prime source of financing enterprises even though they may remain small landholders for a considerable time.

Also I would be a bit cautious about "Soviet aid and advice" as possible large factors in the nationalization process. As you point out earlier, Moscow is not above engaging in plunder where it is feasible; and we know for certain that most, if not all of the smaller "socialist" countries are bitterly critical of Soviet practices. It was certainly one of the elements in the Sino-Soviet rift, whatever the faults of the Chinese in this respect.

25. "Thus the struggle, beginning at the level of national liberation, with the support of the masses moves on past the national level into the permanent revolution. What is essential for the process to develop into the workers state form is the maintenance of the pressures that started the revolution in the first place; a national revolutionary upsurge, a leadership responsive in some measure to the mass pressure, hostility to imperialism concretized in nationalizations, elimination of capitalist elements that oppose this process and the continued pressure of imperialism that prevents any stabilization and adaptation to a peaceful coexistence at some intermediate stage. With this also develops a socialist consciousness which has either been in the background of the movement or has been acquired in the course of the struggle."

As a general description of the revolutionary process as a whole in the colonial and semicolonial spheres, what you say is correct. However, you leave out (although you come to it in your next paragraph) the role of governmental power in this process. For the purposes of our discussion, this is the decisive item.

26. "Does a government that carries out these measures have to be a workers and peasants government?"

I am not sure what you have in mind. Are you thinking of Egypt or some of the other African and Middle Eastern countries? What would be implied is the possibility of other kinds of governments establishing workers states. Even more, that some of these countries are already workers states.

27. "Could it not be carried out by a petit bourgeois government or even a national coalition of anti-imperialist forces at least in the early stages?"

This goes even further despite the saving phrase "at least in the early stages."

28. "The experiences of the twenties and thirties indicated that when the middle class was caught between class pressures the majority invariably gravitated toward capitalism. With the relationships changed in the post World War II period it is more possible for sections of the middle class, when under the pressure of the mass movement, to gravitate towards the working class. This has apparently happened in a few special cases."

In the context of the two previous questions, I am not sure what you have in mind. On the face of it, you seem to be merely describing how much more favorable this aspect of the class struggle is today than in the twenties and thirties. Again on the face of it, you seem to have left out completely the problem at hand—the nature of the governmental power that can establish a workers state in the colonial and semicolonial areas in the absence of a revolutionary Marxist party.

29. "It is clear that the negative features persist in this development. While they follow the main line of permanent revolution the transition is not thorough nor complete. They result in deformed or incompleted workers states and remain on the national level. Where they go beyond the national level it is solely as a measure of national defense. While the leaderships have opposed or eliminated threats from the right they equally oppose tendencies from the left that try to move the struggle to the international level."

This is too general, for we are still left with the problem of tracing developments on the level of governmental power so that these can be properly fitted into the overall context. Because these sentences are so general they lead to speculation as to what countries you have in mind.

These are my reactions to your letter, Bob. I am sorry I could not go into some of the points of special interest in detail. On the other hand, if you would take just one country and check out in detail what happened on the single level of the changes in governmental power, this would advance the discussion considerably. There are various countries on which little has been done in this field as you know. At the convention I suggested that work of this kind on the Chinese revolution might prove fruitful, but I think the same holds true of several other countries where it may be easier to obtain the necessary materials.

Fraternally yours
Joseph Hansen

Appendix IV

On the character of the Algerian government

The following statement by the United Secretariat of the Fourth International was published in the February 21, 1964 issue of World Outlook, *the predecessor of* Intercontinental Press.

For some time the course of the new regime in Algeria has shown that it is a "Workers and Peasants Government" of the kind considered by the Communist International in its early days as likely to appear, and referred to in the Transitional program of the Fourth International, as a possible forerunner of a workers state.

Such a government is characterized by the displacement of the bourgeoisie in political power, the transfer of armed power from the bourgeoisie to the popular masses, and the initiation of far-reaching measures in property relations. The logical outcome of such a course is the establishment of a workers state; but, without a revolutionary Marxist party, this is not guaranteed. In the early days of the Communist International it was held to be excluded in the absence of a revolutionary Marxist party. Experience has shown, however, that this conclusion must be modified in the colonial world due to the extreme decay of capitalism and the effect of the existence of the Soviet Union and a series of workers states in the world today.

An essentially bourgeois state apparatus was bequeathed to Algeria. A crisis in the leadership of the FLN [Front de Libération Nationale] came to a head July 1, 1962, ending after a few days in the establishment of a *de facto* coalition government in which Ferhat Abbas and Ben Bella represented the two opposing wings of neocolonialism and popular revolution. The struggle between these two tendencies within the coalition ended in the reinforcement of the Ben Bella wing, the promulgation of the decrees of March 1963 and the ouster successively of Khider, Ferhat Abbas and other bourgeois leaders although some rightist elements still remain in the government. These changes marked the end of the coalition and the establishment of a Workers and Peasants government.

As is characteristic of a Workers and Peasants Government of this kind, the Algerian government has not followed a consistent course. Its general direction, however, has been in opposition to imperialism, to the old colonial structure, to neocolonialism and to bureaucratism. It has reacted with firmness to the initiatives of would-be new bourgeois layers, including armed counterrevolution. Its subjective aims have repeatedly been declared to be the construction of socialism. At the same time its consciousness is limited by its lack of Marxist training and background.

The question that remains to be answered is whether this government can establish a workers state. The movement in this direction is evident and bears many resemblances to the Cuban pattern. A profound agrarian reform has already been carried out, marked by virtual nationalization of the most important areas of arable land. Deep inroads have been made into the old ownership relations in the industrial sector with the establishment of a public and state-controlled sector. Yet to be undertaken are the expropriation of the key oil and mineral sector, the banks and insurance companies, establishment of a monopoly of foreign trade and the inauguration of effective counter measures to the monetary, financial and commercial activities of foreign imperialism.

Among the most heartening signs in Algeria are (1) in foreign policy the establishment of friendly relations with Cuba, Yugoslavia, China, the Soviet Union and other workers states with the possibility this opens up for substantial aid from these sources; (2) the active attitude of the government toward developing the colonial revolution in such areas as Angola and South Africa; (3) within Algeria the establishment of the institution of "self-management." "Self-management" with its already demonstrated importance for the development of workers and peasants democracy offers the brightest opening for the establishment of the institutions of a workers state.

As a whole, Algeria, as we have noted many times, has entered a process of permanent revolution of highly transitional character in which all the basic economic, social and political structures are being shaken up and given new forms. This process is certain to continue. It will be greatly facilitated and strengthened if one of the main problems now on the agenda—the organization of a mass party on a revolutionary Marxist program—is successfully solved.

The appearance of a Workers and Peasants Government in Algeria is concrete evidence of the depth of the revolutionary process occurring there. It is of historic importance not only for Algeria and North Africa but for the whole African continent and the rest of the world.

Appendix V

From Blanqui to Moncada

by Joseph Hansen

The following are comments by Joseph Hansen from the question period following a talk he presented on "Policies of the Cuban leadership" on December 28, 1967, at the New York City branch headquarters of the Socialist Workers Party. Since the remarks were extemporaneous, several dates and other facts about Blanqui's life required correction, which has been done by the editor. The remarks have been edited for grammatical sense.

QUESTION: Would you characterize the Cuban leadership at present as centrist or revolutionary or what?

HANSEN: Well, I think that the Cuban leaders are revolutionary. I don't think that words such as "centrist" mean too much in a context of this character.

I think it's better to look way back to the beginning of the proletariat's revolutionary struggle—back to 1839, for example. If you go back to that historic occasion, you'll run across the name of a person named Auguste Blanqui who led an attempted *coup d'etat* in France. He got together 500 men, who organized very secretly in a group called the Society of the Seasons, and they attempted to take over the government in Paris—kind of a Fort Moncada like the Cubans carried out in 1953. They held out a couple of days, but then they were defeated. Blanqui was given a death sentence, which was commuted to life imprisonment. He finally got out of prison right before the 1848 revolution in France, but he soon ended up back in prison for participating in an unsuccessful mass popular demonstration that declared a revolutionary government.

There is no question at all that Blanqui was a revolutionary. There is no question whatsoever about the dedication of the people who were involved in the 1839 uprising and so on; these were the first Fort Moncadas.

Blanqui's theory was very simple. His theory was that if a group of revolutionaries were completely dedicated, if they were willing to give their lives for the cause, if they organized in such a way that they were not caught by the government, then they could plot an attack of such a surprise character that it could catch the government unawares. This would then electrify the masses. They would be galvanized by this heroic action and pour into the streets. Then you'd have the power—just like that. That's the way he figured it.

What actually happened when this occurred, even if the workers were for what Blanqui was fighting for, is that they didn't know what it was all about. Some kind of a shooting, some kind of a fracas was going on somewhere. They didn't know quite what it was. It was all over so quickly that it was too late for them to do anything about it. The cops were going all over, shooting up the town. There was nothing the workers could do about it. So the thing didn't work.

Blanqui spent a total of 33 of his 75 years in prison. He was in and out of prison all his life, and some of his stretches were very long. By an historic accident, he was put back in prison just one day before the Paris Commune was established in 1871. The Commune elected him as its president, but the French government wouldn't release him from jail. The opinion of some people was that if Auguste Blanqui had been outside prison at the time of the Paris Commune, given his experience and the amount of thinking he'd given to problems of insurrection and power, the Paris Commune could very well have been successful.

Marx had a very high opinion of Blanqui. He called Blanqui the "brains and heart of the proletarian party in France"—that was Marx's opinion. And the French proletariat, the French workers considered him almost like a saint. When he died in 1881, over 200,000 workers turned out for his funeral.

Revolutionaries since then have had tremendous respect for Auguste Blanqui. One of those who had the greatest respect for him was none other than Lenin. Lenin thought that Blanqui had a good idea in seeing the need to organize an insurrection. Where Blanqui went wrong was in thinking that this would somehow galvanize the masses. You've got to organize the masses prior to such an action, Lenin said.

Blanqui himself came to recognize this after he'd been in prison for some years. He opposed his earlier ideas among his own followers. He said he'd changed his mind, he didn't think it would work, and that they needed to do something else. But he couldn't convince them, he was in a minority. And so they pulled a couple of more "Fort Moncadas," and back into prison he went. A total of 33 years for this heroic figure.

Now, would you characterize Blanqui as a centrist? How would you characterize him? He was a revolutionary.

I would say the same thing about the Cubans. The remarkable thing about the Cuban revolution is that a concept of Blanqui's that was considered to have been completely outmoded by history, and especially by the October revolution, could succeed in our day and age. It took very special circumstances for such a concept to succeed; that's obvious.

One of the factors that made this possible was that the Communist Party in Cuba had acted as a big obstacle to the revolution for a very long time. Things had reached the point where it was so rotten, while at the same time the conditions for the revolution were so ripe, that even this costly tactic could succeed.

My point is that it's not likely to succeed in other parts of the world, or if it does, it will be a way of achieving revolution that has all the odds stacked against it. Even if it could succeed, it's a very costly way. It's not the most effective way.

But I would not want to say that the Cubans are centrist or not centrist. I'd say they are revolutionists who are trying to grasp reality and find their way along the real lines of the class struggle as they see it.

Unfortunately, they are handicapped by the fact that all the concepts of the October 1917 revolution (which by now should have become a primer for the youth of the entire world, just as they learn arithmetic) have been covered over by years and decades of Stalinist slander and misrepresentation. As a result, in many parts of the world comrades have to start from scratch. They start as revolutionists, and they have to find their way painfully and cover ground that has already been covered. This is the reality. The Cubans covered ground that had already been covered, and in the process they won a victory. That's the score.

Appendix VI

Resolution on workers and peasants government adopted at the June 1923 ECCI plenum

The following resolution was adopted at the June 1923 plenum of the Executive Committee of the Communist International.

The relationship between the working class and the peasantry comprises one of the most fundamental problems of the international proletarian revolution. A correct estimate of the relations of these two basic classes of the working population will determine the success of our struggle both prior and subsequent to the conquest of power.

A general estimate of the relation of the proletariat and the peasantry was given in exhaustive detail in the resolution of the Second World Congress of the Communist International on the agrarian problem. It ran as follows:

> I. The urban and industrial proletariat alone, led by the Communist Party, can liberate the toiling masses of the rural localities from the oppression of capitalism and landlords and from the destruction of imperialist wars, which must inevitably recur as long as the capitalist order prevails. There is no salvation for the labouring rural masses except in alliance with the communist proletariat and the ungrudging support of the latter in its revolutionary struggle for the overthrow of the yoke of the landowner and the bourgeoisie. On the other hand, the industrial workers will not be capable of fulfilling their historic world mission of liberating mankind from the oppression of capitalism and from wars, if they confine themselves to narrow, craft and trade union interests and selfishly struggle for the improvement of their own often quite tolerable petty bourgeois position. This is the case with the "labor aristocracy" which comprises the basis of the would-be socialist parties of the Second International but which is in fact the worst enemy of socialism and its betrayer; petty bourgeois chauvinists and agents of the bourgeoisie within the working class movement. The proletariat will be the real socialist class if it comes forward as the vanguard of all the toilers and exploited, as their leader in the struggle for the overthrow of the exploiters. This is impossible unless the class struggle is carried into the countryside, unless the toiling masses of the villages are rallied around the Communist Party of the urban proletariat, and unless the former are educated by the latter.
>
> II. The toiling and exploited rural masses who must be brought into the struggle or at any rate brought over to the side of the town proletariat, consist in all capitalist countries of the following sections:
>
> 1. The agricultural proletariat, wage workers (yearly, seasonal, or daily) acquiring their livelihood by working for wages in capitalist agricultural enterprises, and the industrial enterprises connected with them. The independent organisation (political, military, trade union, cooperative, educational, etc.) of this class (including foresters, artisans on estates, etc.) apart from all other classes; active propaganda and education amongst this class, and the securing of its participation in the Soviet power and the dictatorship of the proletariat is one of the fundamental tasks of the Communist Parties of all countries.
>
> 2. The semi-proletariat or semi-peasants, i.e., those who earn their livelihood partly as wage workers in agricultural or industrial capitalist enterprises, partly by working their own rented plots of land, which afford them only a part of the foodstuffs necessary for the support of their families. This section of the agricultural population is extremely numerous in all capitalist countries. Its existence and peculiar situation is concealed by the representatives of the bourgeoisie and the "social-

ists" of the Second International, who lump them with the general mass of the "peasantry," acting partly with the conscious intention of deceiving the workers and partly under the influence of the blind routine, petty bourgeois point of view. This bourgeois method of gulling the workers is most observable in Germany and in France, but also in America and other countries. If the work of the Communist Party is properly conducted, this section will become its ally, for the position of these semi-proletarians is extremely hard and the gain to be got by them from the Soviet power and the dictatorship of the proletariat is great and immediate.

In certain countries there is no hard and fast line of demarcation between the first and the second groups. Therefore, in certain circumstances, it is possible to organise them together.

3. The petty peasantry, i.e., small agriculturalists, cultivating their own or rented small pieces of land, sufficient to cover the needs of their households, and not necessitating the employment of wage labour. This section would certainly gain by the victory of the proletariat.

Taken together, the above-mentioned sections comprise in all countries the majority of the rural population. Therefore the final success of the proletarian revolution is guaranteed not only in the town, but also in the countryside.

The Fourth World Congress of the Communist International developed and supplemented the resolution of the Second Congress, giving the outline of the agrarian "program of action" (minimum programme) of the Communist International on the agrarian question.

The Second Congress of the Comintern drew up the fundamental postulates for a program on the relations between the working class and the peasantry. The Fourth Congress gave concrete form to these postulates. The present Enlarged Executive meeting of the Communist International must now produce a concentrated *political* formula, which will enable us with the greatest chances of success to carry into practice the decisions of the Second and Third Congresses.

This political formula is—"The Workers' and Peasants' Government."

Since the first world imperialist war, the peasantry has not been what it was before the war. In most countries which took part in the war, large sections of the peasantry have accumulated a certain amount of political experience. As a result there are as observed in recent years serious attempts to create peasant parties which are endeavouring to play an independent political role.

The repeated attempts during recent years to form a Green Peasant International are worthy of note.

On the whole the attempts of the peasantry to conduct a middle policy between the bourgeoisie and the proletariat necessarily failed. In the most advanced bourgeois countries, the bourgeoisie and large landowners continue as formerly to lead the peasantry by the nose. Even where apparently independent peasant parties exist they are led by elements foreign to the peasantry (priests, lawyers, landowners). The labouring peasants are then only the instruments and political cannon fodder of the worst enemies of their class. This is one of the supports of the bourgeois regime. Modern history gives numerous examples illustrating the truth that the wide masses of the labouring peasantry can only defend their political interests in close alliance with the revolutionary proletariat and only on the condition that they give their support to the revolutionary part of the proletariat.

Meanwhile the attitude of the parties of the Second International to the peasantry is changing. The former traditional neglect of the peasantry is giving place to attempts to draw it into the path of counterrevolutionary Social-democratic policy. In proportion as the most important Social-democratic Parties lose important positions in the working class movement and spasmodically seek a new social basis, they inevitably turn towards the countryside and direct their attention to the wealthiest sections of the peasantry.

The task of the Communists is immediately to seize the positions vacated by the Social-democrats within the working class, and while continuing their attacks above all in this sphere, to endeavour to foil the Social-Democrats who are seeking for a new social basis in the countryside, and thus to rally around our banner the agricultural proletariat, and the rural semi-proletariat, and to induce the peasantry to enter into alliances with the revolutionary proletariat.

The mere fact that the Communist Parties adopt and agitate for the slogan of a Workers' and Peasants' Government on an international scale will be sufficient to begin the neutralisation of sections of the medium peasantry and the winning of the petty peasantry over to our side.

The Executive Committee of the Communist International asserts that the great majority of the sections of the International have hitherto displayed an attitude which is extraordinarily inert and extremely harmful to our cause on the question of work in rural districts. This attitude is due, firstly, to the unhappy traditions of the Second International, out of which the most important parties of the Communist International sprang; secondly, to an incorrect theoretical attitude towards the peasantry which professes that from an "orthodox Marxian" standpoint the party of the workers has no connection with the peasantry; and thirdly, to a narrow craft conception of the class struggle of the proletariat. It is the duty of the Communist Parties at the present time to break once and for all with this craft point of view. The Communist Parties must not regard themselves as the parties of the extreme proletarian opposition *within* bourgeois society, as was the case during the period of the development of the Second International. The Communist Parties must develop in themselves the psychology of parties which sooner or later will lead the toiling masses into the fight against bourgeois society, to overthrow the bourgeoisie and to replace it as the rulers of the State. The narrow craft psychology must be replaced by the psychology of parties which possess the will to power, which embody the interest of class hegemony in the revolution. A Communist Party must prepare itself to defeat the bourgeoisie tomorrow and *therefore* today adopt aims common to all the people. It must therefore attempt to attract to the support of the proletariat all those sections which because of their social position, will be able at the critical moment to support the proletarian revolution in one way or another.

The motto of the "Workers' and Peasants' Government," like that of the Workers' Government in its time, does not in any way replace or put in the background the agitation for the dictatorship of the proletariat—the foundation of foundations of Communist tactics. On the contrary, the motto of the Workers' and Peasants' Government, by extending the basis of the tactic of the united front—the only correct tactic for the present epoch—is the path to the dictatorship of the proletariat.

The correct interpretation of the motto of the Workers' and Peasants' Government will permit the Communists not only to mobilise the proletarian masses of the towns, but also to create valuable points of support in the countryside and thus prepare the ground for the seizure of power.

The slogan of a "Workers' and Peasants' Government" will render good service to the communist parties even after the seizure of power by the proletariat; for it will remind the proletariat of the necessity to harmonise its movements with the sentiments of the peasantry in their respective countries, to establish a correct coordination between the victorious proletariat and the peasantry, and to observe a rational policy in the gradual introduction of the economic measures of the proletariat, such as was arrived at by the victorious proletariat of Russia in that period of the Russian revolution which is called the new economic policy.

It will, of course, be understood that the agitation, carried on under the slogan Workers' and Peasants' Government must be adapted to the conditions prevailing in each country, for instance in the United States it will apply to working farmers.

The defence of the economic interests of the peasantry in the spirit of the programme laid down in the decisions of the Second and Fourth World Congresses of the Communist International, must be the starting point for all our agitation for a Workers' and Peasants' Government. The Enlarged Executive therefore demands of the national parties to prepare immediately a concrete program on their relations to the peasantry and introduce corresponding bills in Parliament through their parliamentary fractions. Such bills will be of the greatest possible importance if they really meet the actual needs of the working peasantry, and if signatures in support are collected in the countryside.

As a propaganda slogan, which makes it possible

for us to express in arithmetical form that which, hitherto, has only been expressed in algebraical form, the slogan of a "Workers' and Peasants' Government" will be of the greatest significance in such countries as France, Germany, Italy, the Balkans, Czecho-Slovakia, Poland, Finland, etc. At any rate the victory of the proletarian revolution and its consolidation will nowhere be possible without some assistance from the peasantry. In this sense the slogan of a "Workers' and Peasants' Government" must be the general slogan of the Communist Parties!

While advancing the slogan of a Workers' and Peasants' Government with every insistence, the Executive of the Communist International recommends the Communist Parties not to forget the dangers which will arise from its incorrect application. Both the tactic of the United Front in general and the slogans of a Workers' Government and a Workers' and Peasants' Government in particular, are undoubtedly pregnant with serious political dangers if our parties are not capable of applying them in a revolutionary Marxian spirit. The two greatest dangers connected with the slogan of a Workers' and Peasants' Government are the following:

1. In parties which have not passed through a true Marxian school, the danger arises of interpreting this slogan in the spirit of the Russian S.R.'s, i.e., in the spirit of petty bourgeois "socialism," which regards the whole of the peasantry as one homogeneous class, and which closes its eyes to the fact that different sections exist within the peasantry. The Executive Committee of the Communist International draws attention to the appropriate point in the programme resolution of the Second World Congress, which says,

> the large peasantry is composed of the capitalists in agriculture, who as a rule work their estates with the aid of hired workers, and who are connected with the peasantry only by their low cultural level, their method of life, and their personal manual labour on their farms. This very numerous section of the bourgeoisie is a decided enemy of the revolutionary proletariat. In the work of the Communist Parties in the countryside, the chief attention should be directed to the fight with these sections for the emancipation of the labouring and exploited majority of the agricultural population from the intellectual and political influence of these exploiters.

2. The second danger that insufficiently experienced communists, from the political point of view, may attempt to replace mass revolutionary work amongst the lower sections of the working peasantry by parliamentary combinations, based on no principles, with the so-called "representatives" of the peasantry which often are the most reactionary elements of the bourgeoisie.

While taking these and similar dangers connected with the application of the slogan of a Workers' and Peasants' Government into account, the communist parties cannot, however, abandon the advantages of manoeuvering tactics and must learn to combine the tactics of penetrating right into the very heart of the wide masses with the principles of revolutionary Marxism.

It is self-understood that penetration into the heart of the masses of the peasantry and the motto of "Workers' and Peasants' Government" by no means imply the conversion of our Party from the workers' party into a "Party of Labour" or a "Workers' and Peasants' Party." Our Party, as far as its composition and aims are concerned, must remain a party of the working class, but of a working class which draws in its wake all sections of the working population and leads them into the fight against capitalism.

One of the most important prerequisites for the carrying out of the slogan of the Workers' and Peasants' Government among the rural population is that Communists with particular energy conduct work in the agricultural labourers' unions. In the immediate future, the Communists must exert all their efforts to obtain a majority in the existing agricultural labourers' unions and to organise such unions where they do not exist. The function that agricultural labourers' unions must perform in addition to their other tasks, is to carry out the task of first-class political importance, of carrying the slogan of the Workers' and Peasants' Government among the masses of the peasantry. In this sense, the agricultural labourers' unions must, as it were, serve as a bridge between the Communist Parties and the countryside.

Under no circumstances, however, must the Communist Parties leave these tasks to the agricultural labourers' unions alone. It is one of the most urgent duties of all the parties energetically to conduct the work of gaining the peasant masses for an alliance with the revolutionary proletariat.

Appendix VII

Zinoviev's report at the June 1923 ECCI plenum

The following are excerpts from the report by Zinoviev on the workers and peasants government adopted at the June 1923 plenum of the Executive Committee of the Communist International.

... If you want to understand the psychological character of our parties, in which nihilism dominates where national questions are concerned, you have to consider the psychology of those parties which do not yet feel that they are parties which are aiming and fighting for power and will have to assume leadership in their state. Most of our parties still have the psychology of a mere opposition party of workers within the framework of a bourgeois society, not of a party which feels itself to be the leading power, the bearer of that hegemony which should win the majority of the people, overthrow the bourgeoisie, and supplant that class in its leadership role. All this is very understandable, since our parties in a majority of countries are still extremely weak. Actually the psychology of many of our sections reminds one of the psychology of the best social democratic parties of the Second International: they are filled with a narrow "class" ideology which is close to a guild ideology. They are not yet parties which hope in a relatively short time to defeat the bourgeoisie and to take over the leadership of their countries. This explains why in a number of countries our parties fail to see the importance of the national question. If, for instance, our Yugoslav party already held the opinion that it would tomorrow, if not today, overthrow the bourgeoisie and put itself in its place, it could never claim that the national question was none of its concern. It would, rather, know that in modern Yugoslavia the national problem is one of the most important levers in our hands for the overthrow of the current regime. We need parties which understand not only how to fight for the eight-hour day, but that also understand how to organize the workers in order to fight for the conquest of the majority of the working masses. ...

... Our parties must definitely remain worker parties, but these worker parties must know how to give the right answers to the national question in all those countries where it is a crucial issue.

This is even more true for the question of the peasantry. In this connection we have been especially neglectful and must thus make special amends. This area makes it very clear that we of the Third International still have loose ends left over from the Second International. ...

Our communist parties in many countries exhibit a certain helplessness, even in those countries where the agricultural question is among the most urgent problems and is absolutely decisive. How are these blunders possible? I have already pointed out that we suffer here from the tradition of the Second International. We should not be ashamed to admit this. Not one of us fell from heaven as a finished communist; we all came from the womb of the Second International and its cursed past is our burden. We must try to get rid of that burden as fast as possible.

From all this, comrades, I deduce that the best means to get rid of all these remnants as fast as possible will be to broaden the watchword "worker government" to "worker and peasant government." You will recall the chronology of this. First we had the tactics of the united front and then came the workers' government. Now, I believe, is the time to generally expand this formula....

... We know very well that peasant parties are not able to play an independent political role for very long. The peasantry either follows the bourgeoisie or follows the proletariat. It is our task to do all we can to bring about the latter. We must carefully follow the disintegration taking place within the ranks of the peasantry. We must not remain exclusively an urban party, if we want to defeat the bourgeoisie; we must become the party of the urban proletariat which, at the same time, remains in close touch with the village. The Russian Communist party was a purely urban party for at least two decades. It was capable of defeating the bourgeoisie only after it had established close ties with the peasantry through millions of soldiers and peasants....

We must be clear about the relationship between the watchword "worker and peasant government" and our old formula of the dictatorship of the proletariat. Undoubtedly there are comrades among us who ask, if we now adhere to the watchword "worker and peasant government," do we not give up our goal of the dictatorship of the proletariat? Do we remain, as we were in the past, a workers' party, or do we now become a worker and peasant party?

He who has at all understood what is meant by the tactics of the united front, he who has begun to comprehend what is meant by class-political strategy of the proletariat, must understand that the formula of "worker and peasant government" is the road to the dictatorship of the proletariat and does not mean to negate that dictatorship. A government of workers and peasants in the strictly scientific sense of the word can hardly be realized. The Soviet government is indeed a workers' government. Power is exercised by the working class and its party. The helm of the state is in the hands of the proletariat. But the proletariat and its party know that one must meet the peasantry halfway and let it participate in the government of the country. In brief, the proletariat wants to govern the country intelligently. For this reason the Russian proletariat, taking into consideration the real balance of strength in the country, has been able to persuade the peasants to cooperate and to create a relationship in which the peasant supports the workers. Thus the experience of the greatest revolution, the Russian Revolution, has proven that this is possible. Our communist parties must now profit from the experiences of the Russian Revolution and adapt them to the concrete conditions in each country. Thus, when we develop the watchword "worker and peasant government," it does not mean that we give up the idea of the dictatorship of the proletariat. We cannot depart a single step from this idea. There is no other way toward the liberation of mankind from the yoke of capitalism than the road of the dictatorship of the proletariat, and there can be no other. The only truly and consistently revolutionary class is the working class. But this class, that is, its party, can behave intelligently or stupidly. In this way [worker and peasant government] we will reach our goal considerably more quickly and with fewer sacrifices. We will partially neutralize, partially win to our side, important strata of the peasantry and of the *petite bourgeoisie*. But if we act clumsily, if we regard the great class tasks of the liberation of the proletariat in too narrow a guild sense, we ourselves postpone the moment of victory....

And one more thing. At the Fourth World Congress we explained to you, why in our opinion the New Economic Policy of the Soviet government is an international phenomenon and not merely an episode in the Russian Revolution. We have proved to you that almost every country after the revolution will have to go through a more or less long phase of this type of policy. We all agreed that the New Economic Policy of Soviet Russia is not a Russian phenomenon and that the victorious proletariat of any country will have to face the problem of the appropriate unification of the working class and the peasantry when the time comes. If this is so, and we do not doubt it, the logical conclusion to be drawn is the necessity of a worker and peasant government. If we consider conditions in a number of countries, we do not see a single country for which this solution would not be the most fitting. We now say to the backward workers and peasants: we want to destroy the state of the rich, and we want to create a state of workers. Let us decide to add: for this reason we suggest the formation of a worker and peasant government. If we make a decision along these lines, it will become impossible for the Social-Democratic party to outstrip us even in the parliamentary arena....

Appendix VIII

The social transformations in Eastern Europe, China, and Cuba

In a report to the twenty-third national convention of the Socialist Workers Party in August 1969, Joseph Hansen presented a general review of the characteristics of the post–World War II overturns in property relations, with particular attention to the case of China. The following is an excerpt from that talk.

At the time of the victory of the Chinese Revolution over Chiang Kai-shek and his imperialist backers, our movement was confronted with the necessity to explain the contradiction between certain long-held theoretical postulates and the actual course of events. The postulates were as follows:

1. The peasantry as a class cannot lead a revolutionary struggle through to a successful conclusion.
2. This can be achieved only by the proletariat.
3. The proletariat cannot do it except by organizing a revolutionary Marxist party.
4. Stalinism does not represent revolutionary Marxism; in essence it is counterrevolutionary.
5. Stalinism represents a temporary retrogression in the first workers state; the advance of the revolution will doom it and it will not reappear.

Despite these postulates, which appeared to have been thoroughly established by both weighty theoretical considerations and a mountain of empirical evidence, in the Chinese Revolution the proletariat did not play a leading role as a class. Instead, this role was assumed by the peasantry.

Moreover, no revolutionary Marxist party was formed on a mass scale. Instead, a Stalinist party stood at the head of the revolutionary forces and came to power in a struggle that ultimately toppled capitalism.

Finally, Stalinism was quite consciously cultivated by the new regime. Today this school of thought has culminated in a cult of the personality that if anything has outdone its model in the Soviet Union.

The problem that faced our movement was to explain these contradictions and to determine what lessons should be drawn and what they portended for the future.

So far as the *political* positions of the world Trotskyist movement were concerned, no problem existed. Without exception our positions were correct, ranging from full support to China, despite Chiang Kai-shek, in the struggle against Japanese imperialism to full support for the revolution against Chinese capitalism and the vestiges of feudalism despite the Stalinist nature of the leadership that was thrown to the forefront.

It is very important to remember this, for it constitutes the most positive kind of proof that our movement is a dynamic political formation and not a church dedicated to maintaining the purity of a set of dogmas. One can feel proud in reading the political platforms presented in the documents of that time. They were very good, standing up remarkably well under the test of events.

Problem of the proletarian content

As to the attempts to find solutions to the contradictions between the reality and our theoretical postulates, some of these were clearly in error from the beginning. Others have not held up, or only created fresh difficulties.

In the main, the attempted solutions centered around locating the proletarian content which it was felt must lie at the heart of the Chinese Revolution despite its strange forms and the role of Stalinism.

For instance, in the case of the peasantry, there was speculation that perhaps its true nature had been misjudged. Unlike the peasants of Western Europe and elsewhere, perhaps the Chinese peasants had achieved a proletarian or even socialist consciousness either because of the peculiarities of China's historic background or because of the impact of imperialism on the country.

A current example of this line of thought is to be found in Comrade Moreno's contribution in *Fifty Years of World Revolution*.

Much greater attention was paid to the nature of the Chinese Communist Party. This was only natural since our movement from its very inception has considered the question of the party to be primordial in the process of bringing a revolution to victory. Thus it appeared that the key to the success in China must be sought in the nature of the Chinese Communist Party.

One line of speculation was that Trotsky had made a mistake in concluding that the Chinese Communist Party under Mao had become a peasant party.

Another was that if Trotsky had been right in his conclusion at the time, then it must have changed back into a proletarian organization.

Comrade Morris Stein argued, for instance, if I recall correctly, that there was a steady flow of workers from the cities who went into the countryside and joined the Chinese Communist Party. Their influence, he thought, was sufficient to give a proletarian character to the party.

Another line of speculation concerned the personal qualities and influence of Mao Tse-tung. Some comrades felt that despite everything, when Mao Tse-tung was faced by the supreme test, he had adhered in practice, if not in program, propaganda, or diplomacy, to revolutionary Marxism.

Still another variant was that the very Stalinism of the Chinese Communist Party gave it a proletarian character. The line of thought here was that Stalinism is connected with the workers state in the Soviet Union and that this association therefore makes it proletarian.

At bottom, this view represents an *identification* of Stalinism with the workers state. It is quite a change from Trotsky's position that Stalinism stands in *contradiction* to the workers state, that it is a cancerous growth. As against the *proletarian* tendency represented by Leninism and the Left Opposition, Trotsky considered Stalinism to be *petty-bourgeois* in nature.

Another line of thought, flowing in the same general channel of trying to find something proletarian about the Chinese Communist Party, was the view that this party changed from a peasant party to a "centrist" party, then a "left centrist" party, then an "opportunist workers party," and finally a "workers party."

In the current discussion, the view that Mao's policies should be designated as "bureaucratic centrism" may fall within this frame.

While I am on the point, I should like to say that I fail to see what is gained by this nomenclature.

If we ask what is the class nature of "centrism," whatever its variety, we are compelled to say that it is petty-bourgeois. That is also the class nature of Stalinism. It is petty-bourgeois.

Thus the introduction of the general term "centrism" does not help in answering whether a Stalinist party can become a revolutionary party. It merely suggests a succession of stages in which the class essence of the gradation or series of steps remains obscure.

Marcy, Swabeck, Posadas, and Healy

It was quite clear from the beginning that all these tentative answers to the central problem carried implications that could prove quite dangerous politically; and we were soon to experience repercussions in our ranks. I will mention some of them.

Sam Marcy and his group rapidly came to the conclusion that Stalinism in power equals a workers state. Since a Stalinist party had gained power in China, this signified that a workers state had been established.

From this position, Marcy evolved into a Maoist of such fervor that he was capable of swallowing even the new constitution, announced at the Ninth Congress of the Chinese Communist Party, designating Lin Piao as Mao's heir.

The consistency with which the Marcyites identify Stalinism with a workers state was shown in the most striking way during the Hungarian uprising when they offered critical support to Khrushchev in using Soviet tanks and troops to crush the proletarian rebellion.

The Marcyites adopted the same position in relation to the current invasion and occupation of Czechoslovakia. They even went so far as to help the Kremlin in its efforts to find a propagandistic cover for crushing the upsurge that was pointing in the direction of a political revolution in Czechoslovakia.

Later in the SWP, we had the sad case of Arne Swabeck, one of the founders of the American Trotskyist movement, who proceeded from the theoretical position that only a revolutionary Marxist party can lead a successful revolution. Inasmuch as the Chinese Revolution was successful, he concluded that the Chinese Communist Party must have been a revolutionary Marxist party, and he ended up as a Maoist.

Juan Posadas followed a similar line of thinking, but with an odd twist. Because of Mao's supposed receptivity to genuine Marxism, Posadas came to believe that Mao derived his finest thought from reading the speeches and writings of J. Posadas. Just how this was accomplished was never made quite clear. Perhaps Posadas believed that Mao had set up a Latin American Bureau in Peking that occupied itself with translating Juanposadas Thought into Chinese ideograms so that Chairman Mao could imbibe at this fountain.

The identification of Stalinism with a workers state took a different and perhaps still more remarkable twist in the thinking of Gerry Healy. He maintains that there are two, and only two, roads to a workers state—either under the leadership of a Trotskyist party or under the leadership of a Stalinist party.

Thus in the case of Cuba, Gerry Healy refuses to recognize the existence of a workers state because the revolution was headed by neither a Trotskyist party nor a Stalinist party.

Wohlforth lays it on the line

If you wish proof of this aberration, it has conveniently been made available in the most recent issue of the *Bulletin* (August 26). On pages S-5 and S-6, Tim Wohlforth, who seems to have displaced Cliff Slaughter as Healy's chief apologist, explains this remarkable theory.

In Eastern Europe, he says,

> The very process of expropriation of capital in these countries was accompanied by a process of the creation of this workers' bureaucracy through the taking over of the government by a workers' party, the Communist Party, and the purging of the government of all forces unreliable to the tasks this party had to carry out—some positive social tasks as well as reactionary tasks.
>
> The Castro government is in no sense a workers' bureaucracy. In fact Castro has carried out a series of purges against even Stalinist elements within his government—as illustrated by the two Escalante affairs—and maintains complete control in the hands of the petty-bourgeois nationalist forces who came to power with him.

> [Then Wohlforth gets down to the nitty-gritty:] In Cuba, and only in Cuba, the nationalizations were not accompanied by the emergence of a government controlled by the Stalinists.

We hardly need any further enlightenment from this Healyite theoretician. His position is that if the process that actually occurred in Cuba had been led by a Stalinist, say Blas Roca or Aníbal Escalante, then the Healyites would have at once agreed that a workers state had been established. If Blas Roca or Aníbal Escalante had purged Fidel Castro and Che Guevara this would have been proof positive.

But since the Stalinists in Cuba were outflanked and bypassed from the left by fresh revolutionary forces, the Healyites find it incompatible with their dogma to admit that a workers state has been established there.

It is this reactionary theory that has led the Healyites, out of concern for consistency, to commit such abominations as to call Castro another "Batista," to offer critical support to Cuban Stalinism when Castro became alarmed at the growth of bureaucratism, and to speculate, as they did openly in their press after Che Guevara left Havana in 1965 for another "assignment," that Castro had murdered his comrade-in-arms.

Now for the icing on the cake. The Healyites make a great show in their press of alertness to the danger of succumbing to Stalinism. However, they have not set a very good example in practice. Besides succumbing to the temptations of Stalinism in Cuba, they succumbed in China.

During the "cultural revolution," the *Newsletter* suddenly blossomed with rave articles about Mao's Red Guards. It was quite a sight to see the great red banner of Maoism lifted high in the *Newsletter*. This lasted but a short time. Praise for Mao's Red Guards vanished as abruptly as it had appeared. For the past two years, the *Newsletter* has hardly mentioned the "cultural revolution."

What happened? No explanation was ever offered. I suppose that the headquarters gang managed to get the ailing author of the articles back into a straitjacket and that was that. It never occurred to them that he was only acting in strict consistency with Gerryhealy Thought.

Four main results of war

The world Trotskyist movement never landed in such blind alleys as the ones in which Marcy, Swabeck, Posadas, and Healy are now to be found. At the same time, I think it is just to say that we have not yet achieved a fully satisfactory unified theory.

Perhaps we are now in position to accomplish this. With good fortune, this may be one of the outcomes of the current discussion.

The method we should follow is that of historical materialism—not the "objectivist" theory, the "accident" theory, or "eclectic dualism." Studies pursued in accordance with the method of historical materialism are the most likely to bring solid results. So let us look at the process that brought into the world the second generation of workers states.

World War II had four main consequences: (1) the victory of the Soviet Union; (2) the weakening of world capitalism as a whole; (3) the resulting temporary strengthening of Stalinism; (4) an upsurge of revolutionary struggles in both the imperialist centers and the colonial areas.

These four results shaped the course of history for some time, above all the advance of the world revolution.

Eastern Europe

In the case of the East European countries that were occupied by the Soviet armies as they moved toward Berlin, the overturn of capitalism in those areas was explainable as a direct consequence of the victory of the Soviet Union over German imperialism.

The armed struggle was carried on by the Soviet armies and the resistance movement operating in conjunction with them. The capitalist governments collapsed as the Soviet troops advanced. They were replaced by governments in which Moscow, standing behind local Stalinist parties, exercised power.

For a time the Kremlin retained the capitalist structures in Eastern Europe, evidently as bargaining pieces in trying to reach some kind of world settlement with Western imperialism.

When this bid was turned down and Washington opened up the Cold War, Stalin responded by destroying the capitalist structures in the countries occupied by the Soviet armies.

Imperialism was too weak to block the overturns. Naturally, there was a great hue and cry. But no capitalist country in Europe had the armed forces required to push back the Soviet armies. Even the U.S. armed forces were disintegrating.

The economic forms that replaced the capitalist structure in Eastern Europe were patterned on the economic forms in the Soviet Union. The structure of the state was likewise based on the Soviet model.

The proletarian element in these newly set up workers states clearly derived from the economic forms that were "structurally assimilated," to use the descriptive phrase applied by the comrades in Europe at the time.

The source of the reactionary Stalinist element, that is, the totalitarian political forms, was the Kremlin bureaucracy, the parasitic ruling caste which was keenly alert to the need to set up a replica of its own formation in these satellite states. Possible sources of political dissidence were handled with frame-up trials and purges.

We, of course, favored the overturns in Eastern Europe although we were absolutely opposed to the means used. To us, the overturns constituted fresh proof that the October Revolution was still alive. Stalin had not succeeded in destroying the foundations of the workers state. Despite himself he had had to export Soviet property forms, if only as a defensive measure against imperialism.

At the same time we were fully aware that the basic policy of the Soviet bureaucracy was "peaceful coexistence" with imperialism and that in accordance with this policy Stalin had once again, during these very same years, betrayed the big revolutionary upsurges in Italy, France, and elsewhere.

Yugoslavia

Let us now consider Yugoslavia. Here again, the Soviet victory was the decisive element. This victory served to inspire the Yugoslav people who had already become armed during their struggle against the German occupation.

The Yugoslav Communist Party had played an auxiliary role in the Soviet military defense by organizing the resistance in Yugoslavia against the German occupation and by pinning down German forces through guerrilla warfare. The armed struggle in Yugoslavia was thus linked to the victories of the Soviet armies.

But the Soviet armies did not play a direct role in Yugoslavia as they did in countries like Bulgaria.

British and American imperialism sought to counter the government set up by Tito by bolstering the forces favoring the monarchy. However, they were too weak to succeed in this, even with the connivance of Stalin. The armed forces under Tito smashed the counterrevolution and became the sole real governing power in Yugoslavia.

This government, in turn, took the steps ending capitalism in Yugoslavia. The economic forms that replaced capitalism were modeled on those in the Soviet Union.

In the political arena, Tito, in true Stalinist style, crushed all dissidence or what might appear to be a potential source of dissidence from the left.

Although the independent role played by the Yugoslav Communist Party under Tito was much greater than that of the Communist parties in countries like Rumania and Czechoslovakia under the Soviet occupation, the basic pattern of the process that ended in the establishment of a deformed workers state in Yugoslavia was the same.

Let us turn now to China. The main condition for the peculiar form which the revolutionary process took there was the same as in the East European countries and Yugoslavia—the victory of the Soviet Union in World War II.

The two other conditions following from this one were likewise the same—the weakening of world capitalism and the temporary strengthening of Stalinism.

As for the revolutionary upsurge touched off by the course of the war and its outcome, this occurred on the colossal scale of the most populous country on earth.

As in Eastern Europe and Yugoslavia, the Soviet armies played a certain role by their proximity in the final stage of the war against the Japanese imperialist aggression, but to a lesser degree than in the European theater.

There were other differences, some of them of an unexpected nature.

China's historic pattern

I should like to suggest that the first of these was the strong resemblance of the opening phases of the third Chinese revolution to the revolutions of former times in Chinese history.

The earlier revolutions followed a cyclical pattern. When the exploiting classes in China reached the point of exerting intolerable oppression on the masses, the entire economic system tended to break down. The remarkable canal system upon which so much of Chinese agriculture depended fell into disrepair. It became increasingly difficult to feed the population. Famines began to occur. The central authority became increasingly hated. Finally, the peasantry, goaded to desperation, began to link up, and, more importantly, to organize for battle.

A phase of armed struggle opened, with its guerrillas, focal centers, and peasant armies. Eventually these armies conquered, and a new government, headed by the leaders of the insurgent armies, came into power.

The new government at once went to work to repair the ravages of the civil war, to reduce the exploitation of the peasants, to divide up the land at the expense of the former landlords. The canal system was rehabilitated and extended, once again assuring a dependable supply of food for the population.

The army hierarchy that constituted the new government naturally soon displayed concern for its own comfort, ease, and even modest luxuries. The hierarchy developed into a privileged bureaucracy. The land became concentrated once again in fewer and fewer hands and the new dynasty came to represent the new landlords. The oppression of the peasantry became worse and worse and the system began to break down once again.

The most interesting part of this ancient pattern is the way the peasants succeeded in uniting and building armies imbued with a central political purpose and capable of smashing the old regime and putting a new and better one in power.

A comparison of this phase of the old pattern with the first stages of the third Chinese revolution would, in my opinion, prove highly instructive.

For one thing, it should help counteract the compulsion felt by our movement for so long to find some kind of proletarian quality in the Chinese peasants to account for their remarkable capacity to create a peasant army imbued with revolutionary political aims.

In any case it would make a very good research project for some young Trotskyist theoretician. So much for that point. We come now to more important items.

New world context

Upon achieving their victory in 1949, the peasant armies of the third Chinese revolution were, of course, confronted by a quite different world from the one their forefathers faced.

First of all, the class nature of the enemy was not the same. In addition they found themselves up against the invading armies of Japanese imperialism, and a little later a fresh threat of invasion from Chiang Kai-shek's American backers, who launched the Korean War and carried their aggression up to the Yalu River.

On top of this, the Chinese peasants established their government in the age of nuclear power, television, jet engines, intercontinental missiles, space rocketry. It was a world dominated by two superpowers, the United States and the Soviet Union—the one tied in with Chiang Kai-shek and standing behind the armies of President Truman and General MacArthur, the other associated with the common struggle against Japan, economic planning, and the immense achievements since 1917 that had lifted Russia out of abysmal backwardness.

Thus the consequence of the victory could not be a mere repetition of China's ancient cycle of revolution and counterrevolution, hinging on the status of agriculture and the private property relations associated with it.

The victory won by the Chinese peasant armies was bound to be shaped by the international context in which it occurred.

Role of armed struggle

The capacity displayed by the Chinese peasants to mobilize themselves in the absence of leadership from the Chinese proletariat gave the armed struggle in China extraordinary force and staying power. Here, too, a special study might provide our movement with very valuable new material.

In checking back in the documents written when China first came up for intensive discussion in our movement, I was struck by the absence of consideration of the role played by the sustained armed struggle.

For instance, in the May 1952 resolution of the International Executive Committee of the Fourth International, which was published in the July–August 1952 issue of *Fourth International,* there is a list of the ways in which the Soviet bureaucracy sought to block the Chinese Revolution from developing into a proletarian revolution. Among the ways, we are told, was the following: "By the pressure exerted upon the Chinese CP to maintain the tactic of guerrilla warfare, and not to attack the big cities."

This could be taken to mean that Stalin favored rural guerrilla warfare for a prolonged period, but was against urban guerrilla war or, more likely, was against the deployment of the peasant armies to take the big cities when that stage of the guerrilla struggle was reached. At one time, of course, he inspired an opposite course—of attacking cities prematurely.

The resolution contains nothing more than this about the import of the armed struggle in the Chinese Revolution.

It is obvious, I think, that if the 1952 resolution had been written in the light of the Cuban experience, or even in the light of the Algerian experience, that a quite different approach would have been taken on this question.

The truth of it is that quite large forces were involved in the armed struggle even in the early stages. In his successive campaigns to liquidate the so-called soviets set up by Mao in Kiangsi in the early thirties, Chiang Kai-shek utilized armies numbering in the hundreds of thousands.

Three of these massive campaigns were defeated by the revolutionary peasant armies, and in 1931 Mao proclaimed a "Chinese Soviet Republic" in this region. It took two more huge campaigns to dislodge this government and compel Mao to begin the Long March in 1934.

A new base was established in Shensi. For a time the armed struggle against the Chiang Kai-shek government was given up in favor of an alliance with the Chinese bourgeoisie and its political representatives. However, the armed struggle continued for a number of years against the Japanese imperialist forces; and in this struggle the revolutionary peasant armies gained in experience and above all in size until they numbered in the millions. We can well appreciate the pressure they exerted to carry the struggle through to the end.

These armies were highly organized—as was required to defeat the enemy—and thus gave rise to a structure of command with vast ramifications. It would be a great contribution to our knowledge if we could know the absolute size of this network, its relations with other mass organizations, and what changes may have occurred in its outlook after the victory.

Workers and peasants government

The role of the peasant guerrillas and the peasant armies is intimately linked to the role played by the successive governments that were set up in the bases controlled by them.

According to Mao, the government of the Chinese Soviet Republic in Kiangsi had 9,000,000 persons under its rule. In relation to China as a whole that was only a modest number. Just the same it was greater than the population of Cuba today.

In 1937, Mao reduced the "Chinese Soviet Republic" to a "regional authority" covering Shensi, Kansu and Ninghsia. The number of subjects was probably a couple of million at most—say a population something like that in Albania today. Nevertheless from this base, Mao's regional government expanded on a big scale during the war against the Japanese imperialist invaders. Similar regional governments were set up until a hundred million persons or so came under the rule of "Red" or "People's" China.

Thus when the workers and peasants government was established in Peking in 1949, long years of experience in wielding government power had already been accumulated by the apparatus under Mao's command.

How to handle a huge military structure, undertake public works, collect taxes, apply oppressive measures, grant concessions, judge which political currents should be ruthlessly stamped out (such as the Trotskyists) and which should be brought into a "coalition" (such as the "democratic-minded" capitalists and their political parties); how to conduct a foreign policy in keeping with the interests of the apparatus—in short, the whole business of running governmental affairs was already old stuff for the Maoist team.

Thus the workers and peasants government headed by Mao that was established in 1949 had a long background of experience that was invaluable in the task of getting things going and rehabilitating the country after the destruction, dislocations, and havoc China had suffered under Chiang Kai-shek and the imperialist armies of Japan.

In the early years not much attention was paid to the sector of China governed by Mao. Thus it is difficult to form an accurate picture of the way Mao ruled in the period before moving to Peking in 1949 and establishing his fourth capital there. (Juichin, Pao An, Yenan, Peking.)

What kind of justice prevailed under Mao during these decisive years? Was it balanced and fair? Was democracy practiced? Did even a semblance of democracy exist? Or did Mao follow the practices he admired so much in Stalin?

I think that we can make a fairly good guess.

When the peasant armies finally took the cities, they not only put Chiang Kai-shek and his forces to flight, they suppressed every move of the proletariat to engage as an independent force in the revolutionary upsurge. In following this policy, Mao was not initiating something new, he was continuing what he had practiced for years. Stalinism was congenital in the new regime.

Stalinism, a temporary phenomenon

Perhaps this is the place to consider Trotsky's thesis that Stalinism was a temporary phenomenon, doomed to disappear with the advance of the revolution. This is absolutely correct on a historic scale. Trotsky based it on the consideration that with the success of the proletarian revolution in one or more advanced capitalist countries, the standard of living could be raised so rapidly as to destroy Stalinism economically, since Stalinism arose as a product of a backward economy in a country subjected to extreme isolation and pressure by world capitalism.

But Trotsky did not speculate on what might occur if the proletarian revolution in the advanced capitalist countries was delayed for several more decades while the revolution conquered in areas still more backward than Czarist Russia.

We have seen what happens in this case. It is a matter of history. Stalinism is temporarily strengthened and its death agony is prolonged.

Trotsky's thesis nevertheless caused many comrades to scan Maoism with the hope that it might prove to be anti-Stalinist and thus provide early confirmation of Trotsky's prognosis on the historic fate of Stalinism.

Mao's policy in Indonesia and his course in the "cultural revolution" have shown how misplaced these hopes were.

Birth of Chinese workers state

Let us continue with our analysis.

The workers and peasants government that began wielding power in Peking in 1949 was decisive in another respect in shaping the ultimate outcome of the Chinese revolution.

It was this government that finally destroyed the capitalist state and established a workers state in China. This took place despite Mao's "New Democracy" program of maintaining capitalism for a prolonged period. The tasks faced by the new regime, particularly when they were compounded by the aggression of American imperialism in Korea, were of such order that they could be met only through economic forms that are socialist in principle.

The establishment of a workers state in China offered the most striking testimony as to the validity of the basic premise in Trotsky's theory of the permanent revolution; namely, the tendency of revolutions in the backward countries to transcend the bourgeois-democratic phase and turn into socialist revolutions. Our movement has correctly placed a great deal of stress on this; it is not necessary for me to repeat it here.

What I should like to call special attention to is the link in the revolutionary process through which this qualitative leap was made possible—the workers and peasants government.

From the theoretical point of view this is the item of greatest interest, for it was this government that set up the economic forms modeled on those existing in the Soviet Union, repeating what happened in Eastern Europe and Yugoslavia.

The possibility of workers and peasants coming to power had been visualized by the Communist International at the Fourth Congress in 1922. But the Bolsheviks held that such governments, set up by petty-bourgeois parties could not be characterized as proletarian dictatorships, that is, workers states.

The Bolsheviks were firmly convinced that petty-bourgeois parties, even though they went so far as to establish a workers and peasants government, could never move forward to establish a workers state. Only a revolutionary Communist party rooted in the working class on a mass scale so as to be able to lead it into action, could do that.

The experience of China showed that in at least one case history had decreed otherwise.

This came on top of the experience of Yugoslavia and in Eastern Europe where it can be argued that the implications were not so clear cut because of the role played by the Soviet armies, the catastrophe suffered by German imperialism, and the revolutionary crisis suffered by the other capitalist powers in Europe.

It was precisely because of the adjustment that would be required in the hypothesis advanced by the Fourth Congress of the Communist International that our party moved so cautiously and sought to explore every possible alternative before it agreed to recognize that a workers state had been established in China. We take a very serious attitude toward theory.

The thoroughness with which we sought to examine the consequences of the Chinese experience served as good preparation for what happened in Cuba some ten years after the Chinese victory. We were able to follow the pattern of events in Cuba with ease.

The most gratifying aspect of this from the standpoint of theory was that the pattern of the Cuban Revolution decisively confirmed the principal conclusions we had reached with regard to China.

Cuba and Algeria

The key item in Cuba was the workers and peasants government established in 1959 by a petty-bourgeois political force, the July 26 Movement.

As in the case of China, this new Cuban government, which had been brought to power through a hard-fought armed struggle and a revolution of the most deep-going and popular character, could not meet the giant tasks it faced, particularly in face of the violent reaction of U.S. imperialism, without toppling the capitalist structure and establishing economic forms that were socialist in principle.

Once again, these were modeled by and large on those in the Soviet Union. Even more than in the case of China, the very possibility of a workers state in Cuba of any durability hinged on the existence of the Soviet Union. The appearance of a viable workers state in Cuba was thus a consequence, in the final analysis, of the victory of the Soviet Union in World War II.

The pattern was similarly visible in the Algerian Revolution. In this instance, however, no workers state was established. Instead the workers and peasants government was brought down by a military coup d'état in June 1965 after some three years in power.

This was proof that the establishment of a workers and peasants government does not automatically guarantee the subsequent establishment of a workers state.

In the case of Cuba, a significant new development was to be observed. The leadership that came to power, while it was petty-bourgeois, was not trained in the school of Stalinism. It stood to the left of the Cuban Communist Party.

The importance of this cannot be overemphasized. The team headed by Fidel Castro and Che Guevara constituted the first contingent of a new generation of revolutionists that cannot be brainwashed by either Moscow or Peking.

Trend toward classic norm

On the broad scale of the post–World War II period, this constitutes a watershed.

The deformation of the revolutionary process in Eastern Europe, in Yugoslavia, in China, in North Korea and North Vietnam was a resultant of the revolutionary upsurge following World War II coupled with the temporary strengthening of Stalinism.

The expansion of Stalinism, however, intensified its internal contradictions and this led to a series of crises that finally culminated in the Sino-Soviet conflict and the spread of "polycentrism." Stalinism has thus been greatly weakened. Even in its Maoist form, Stalinism now faces an increasingly dim future.

On the other hand, the establishment of a series of workers states as the consequence of successful revolutions has greatly strengthened the world revolution and its perspectives.

This means a growing tendency internationally toward a revolutionary pattern that comes much closer to the classic norm in which the proletariat moves into the foreground. Evidence of this is to be seen in the shifting of the axis of revolutionary struggles in the backward countries from the countryside to the cities. The events in France in May–June 1968 showed what explosive potential now exists in the imperialist centers of the West. The ghetto uprisings in the United States and the upsurge among the student youth internationally have offered further corroboration of the trend.

We can conclude from this that the next revolutionary victory, wherever it comes, will in all likelihood go even further than the Cuban Revolution in departing from the deformation imposed by the pernicious heritage of Stalinism. The Leninist norm, calling for construction of a fully conscious revolutionary-socialist combat party, will acquire full force and validity as revolutionary situations develop in the strongholds of world capitalism.

Consequences

What are the main consequences of viewing the Chinese Revolution along the lines I have indicated so far as the current discussion is concerned?

First of all, I would say that it is much easier to see the role played by the peasantry and its petty-bourgeois leadership. We can call them what they are, *petty-bourgeois,* without seeking to conjure away this fact or to ameliorate it by speculating that after all these forces must have been proletarian in some shape or fashion, otherwise the peasantry and the Stalinized Communist Party could not have played the role they did.

Secondly, we can see much more easily how a proletarian element did finally come into play in the Chinese Revolution through the governmental power that established economic forms modeled on those of the Soviet Union.

Thirdly, we can more easily see the continuous thread of Stalinism in China from the very beginning up to the current stage marked by the crisis and fierce factional struggle of the "cultural revolution." It is not necessary to look for periods in which Stalinism presumably vanished—only to reappear. We eliminate this awkward hypothesis which would require us to explain how Stalinism in China could have died in the flames of a peasant upheaval only to arise again from the ashes of the "great proletarian cultural revolution."

Fourthly, we can much more easily grasp the origins of the bureaucracy in China, how it was shaped by Stalinism as it came into being, and what a substantial element this bureaucracy actually is in the Chinese social and political scene.

Fifthly, we are in better position to understand the interrelationship between Mao's domestic and foreign policies, and particularly in the case of his foreign policy to see how its basic design is to safeguard and advance the position of the bureaucratic ruling caste and why this gives his foreign policy its nationalistic "peaceful coexistence" characteristics and its capacity to alternate between rank opportunism and adventuristic ultraleftism. It becomes easier to see the true origin of Mao's foreign policy and to avoid the error of mistaking the *resultant* of the clash between Peking's policy and the contending policies of other countries with what Mao seeks to achieve.

Sixthly, by considering the pattern of the Chinese Revolution in conjunction with the patterns in Eastern Europe, Yugoslavia, Cuba, Algeria, we can much more readily appreciate the limitations of the lessons to be drawn. It is easier to avoid unwarranted and incorrect extrapolations that could prove very misleading and dangerous.

In mentioning these consequences, I should like to stress that they are derivative. They follow from viewing the Chinese Revolution in the way I have suggested.

What is most important, of course, is to weigh the validity of this analysis of the pattern of the Chinese Revolution and its connection with the patterns in Eastern Europe, Yugoslavia, Cuba, and Algeria.

In any case, as the discussion develops internationally on this subject, the most fruitful contributions may well be those that seek to fill in the extensive gaps that still exist in our knowledge of some of the phases of the Chinese Revolution that are of the greatest interest from the standpoint of theory.

State capitalism

Postscript:

Because of time limitations it was not possible for me to do more at the convention than barely refer during my summary to a point that should be considered logically in conjunction with the question of the degenerated or deformed workers states and their relationship to Stalinism. This is the peculiar state structures of countries like Egypt and Burma.

As is well known, in these countries the government has taken over the bulk of the means of production with the exception of agriculture.

The nationalizations are so extensive, in fact, that quantitatively the situation appears comparable to what exists in the workers states. As a result it is tempting to equate them with workers states; and this has been done—incorrectly so—by various currents.

One procedure of those who make this error is to call them workers states. Another is to call them state capitalist; but—still equating them with workers states—to call countries like the Soviet Union and China "state capitalist."

The essential difference between states like Egypt and genuine workers states is to be found in their different origin. In every instance, the workers states, whether deformed or otherwise, have emerged as products of revolutions. Through armed struggle, through upheavals involving the masses on an immense scale, the people have overthrown their capitalist oppressors, displacing them from power in the most thoroughgoing way.

In countries like Egypt, upheavals on this scale have not occurred. The usual pattern is that a sector of the officer caste takes over, generally through a coup d'état, occasionally ratified through partial mobilization of the masses, who, of course, are in favor of ousting the old regime.

The new government is fearful of the masses. One of the first things it does is to block the masses from mobilizing, at least in a massive revolutionary way. The new government aims at giving capitalism a new lease on life after a period in incubation under auspices of the state apparatus.

The officialdom is thoroughly aware of the ultimate perspective, and conducts itself accordingly. How the state machinery is used to spawn millionaires was graphically demonstrated in Mexico.

It is obvious that the qualitative nature of nationalizations is determined by whether they originate in a thorough-going revolutionary struggle or in measures undertaken by a sector of the officer caste or their political representatives, who may even have in mind forestalling a popular revolution by setting up a simulacrum of a workers state. This phenomenon can be quite correctly placed under the general heading of state capitalism.

What is demonstrated by the extensive nationalizations in countries like Egypt—and the less extensive ones in Mexico and elsewhere in Latin America—is the enormous pressure being exerted on a world scale to bring capitalism to a close and to move into the epoch of socialism. Private capitalism has become so antiquated, so outdated, that capitalist governments everywhere are compelled to intervene more and more extensively in the very management of industry if they hope to prolong the death agony of the system a bit longer.

The growth of state capitalism also testifies to the depth of the crisis in revolutionary leadership observable on an international scale. Prime responsibility for this lies with Stalinism.

The overhead cost of the many betrayals of the most promising revolutionary openings, from Germany in the early thirties to Indonesia three decades later, can be measured, among other ways, by the growth of statism, the direct intervention of the capitalist state in the economic system.

The importance of the occurrence of a *revolution*, as one of the criteria in determining that a workers state has come into existence, is very clear in the case of Cuba.

Because they do not recognize this criterion, the Healyites refuse to acknowledge that a workers state exists in Cuba. They lump Cuba with Egypt, Burma, Syria, and so on.

They are inconsistent in not placing China and Yugoslavia in the same category. They seek to avoid this inconsistency by making the existence of *Stalinism* the decisive criterion. This shows that in the final analysis they are incapable of distinguishing between revolution and counterrevolution.

The qualitative difference that a revolution makes in nationalizations is evident in the difference in durability of the takeovers in countries where a revolution has occurred and countries where it has not occurred.

This is because of the fact that the old ruling class is smashed in the one instance and only temporarily displaced in the other while the state structure is used to rejuvenate the system. The marked difference in popular consciousness is likewise of prime importance.

Cuba and Burma offer striking examples of these differences.

A comparative study along these lines would undoubtedly prove highly instructive.

Appendix IX

From 'Two Proposals'

By George Breitman

The following are excerpts from the article "Two Proposals" by George Breitman in *SWP Discussion Bulletin*, Volume 25, Number 12, 1965.

2. That we change the transitional slogan "For a Workers and Farmers Government" to "For a Workers Government."

The present slogan and the one proposed both are designed as bridges to the idea of a government of the revolutionary workers and their allies among other sections of the population—farmers, minority groups, women, youth, parts of the petty-bourgeoisie, etc. The trouble with the present slogan is that it mentions only one of these potential allies (farmers), and that these are not the most important of the allies, either numerically or socially. Since we can't include all the allies in the slogan, it would be better in my opinion to include none, and to list them all in the explanations we have to make of the slogan; we have to make explanations with both slogans.

There was a time when the farmers were a much bigger section of the population, and when their relative political weight was heavier. Such was the case in 1938 when the present slogan was adopted. But technological change has altered the situation considerably. In 1938 those Americans occupied in agriculture, including all classes and their families, represented between 21 and 22 percent of the population; today the figure is around 8 percent, and the trend continues to be down. It is wrong today to single out the farmers above all the other potential allies of the revolutionary workers when, to take one example, the Negro people are both more numerous (10 to 11 percent) and more dynamic.

In 1938 there were differences and a discussion about the slogan (see articles by Burnham and Weber and conversation with Crux in the August, 1938 bulletin). I hope my proposal will not be connected in any way with that discussion (or be misconstrued as an "underestimation" of the revolutionary potential of the farmers). The proposed change is motivated primarily by the different rank among our potential allies that the farmers have been shifted to by changes in American capitalist society during the last 27 years.

But I would like to call attention to a point made in 1938 by both Crux and Weber in their defense of the present slogan. Crux said, "The farmers play a very important role in the United States. In England, this is not a very important question because the workers are the overwhelming majority." Weber, following this cue, said the workers and farmers government slogan does not apply universally; "It would not at all apply, for example in England. But it does apply in the United States." Because in England the farmers had, in 1938, become "a negligible factor, numerically and economically," forming "families and all, only some seven percent of the population. Thus in England it would be quite unnecessary to argue the point, in a transitional program, of whether to call for a workers government or a workers and farmers government."

Detroit, Michigan
AUGUST 1, 1965

The above two proposals were submitted last January too late to reach the N.C. plenum. They were then mailed to the N.C. members in April, but never evoked any response. They are now submitted for action by the coming convention, under any point on the agenda deemed most suitable, or for referral by convention action to the new N.C. for disposition.

I would add only this: "For a Workers Government" looks a little sparse. Perhaps the formula should read: "For a Workers Government—for a government based on, representing and acting for black and white factory, farm and office workers and their allies." On certain occasions, it could be referred to in shorthand as "A Black and White Workers Government" or "A Factory, Farm and Office Workers Government"; or "a government of factory, farm and office workers and their allies."

Appendix X

A proposed change in transitional slogans

The following motion from the Political Committee was adopted by the 1967 National Convention of the Socialist Workers Party.

The Political Committee recommends that the party change the transitional slogan "For a Workers and Farmers Government" to "For a Workers Government."

Before explaining the reasons for the recommended change, it seems useful to review briefly the background of the two slogans. Both are designed for common use as a bridge to the idea of a revolutionary government of the workers and their allies. This transitional concept was developed by the early Comintern then led by Lenin and Trotsky.

Its purpose was to initiate mass consciousness of the need for class struggle politics, as against the social democratic line of political coalition with the capitalists, and to develop that consciousness to the logical revolutionary conclusions.

In 1938 this Bolshevik propaganda device was adopted by the Fourth International and by the Socialist Workers Party. Under Trotsky's leadership the theses involved were brought up to date by including historic experience with the Stalinist variety of political class collaboration.

The call for a government of the workers and their allies is intended to lead toward mass recognition of the need for a dictatorship of the proletariat, as conceived by the Bolsheviks and as brought into being by the October 1917 revolution in Russia. That class dictatorship had nothing in common with the dictatorial bureaucratic regime that later evolved out of the Stalinist degeneration in the Soviet Union. It represented genuine workers democracy and it was the only effective way in which capitalist rule could be displaced by working class rule. The workers and their allies were armed, the counterrevolutionary capitalists defeated and disarmed. All power was taken into the hands of the toilers through the soviets, spearheaded by the working class under the leadership of the Bolshevik party. Capitalism was abolished and new foundations laid from which to proceed toward the construction of a socialist order.

The slogan for a government of the workers and their allies—counterposed to the false course of crossing class lines in politics and seeking a governmental coalition with capitalists—can get a hearing among militants who have not yet recognized the necessity for a proletarian dictatorship. They can be influenced by the concept of a government by and in the interests of the workers and the masses generally. In embracing that concept they take a forward step, even if they retain illusions that their basic problems can be solved through the electoral process and parliamentary action.

Their struggle for a genuine workers government will teach them that the capitalists won't allow the issue of political rule to be decided peacefully by a simple majority decision. The capitalist class will resist any attempt to break its monopoly over political control of the country which, although dressed in bourgeois-democratic trappings, constitutes an actual class dictatorship. History teaches that—while prepared to use a reformist labor facade as an instrument for its own domination—the capitalist minority will plunge the nation into civil war rather than yield to actual majority rule by the workers and their allies. It follows that under the impact of events illusions about peaceful social change through majority decision will have to give way to preparation for fierce class battles. Consciousness will grow that in order to establish a genuine workers government it is necessary to defeat the capitalists in all-out struggle.

Although we advocate anticapitalist electoral activity, we do not project the concept of parliamentary action alone. Our aim is to use the electoral sphere as a means to advance a platform for mobilization of the masses in all-sided class struggle. As mass radicalization is deepened and class battles grow sharper, the way is opened in turn to project the essential concepts of the dictatorship of the proletariat.

In this longer-range sense the slogan for a government of the workers and their allies becomes a pseudonym for the concept of the proletarian dictatorship. As such it helps to get around prejudices against the concept of a class dictatorship that have arisen due to the hateful image of Stalinist totalitarianism. Minds can be opened to an explanation of the need for the toilers to take the power into their own hands and of the necessary measures toward that end. The class treachery of the misleaders can be exposed and support won for our revolutionary-socialist program.

In considering the question of working class allies relative to the phrasing of transitional slogans the Bolsheviks gave special prominence to the peasants. Economically the peasants represent a survival of the productive system under feudalism. They are not a social class but a series of layers of social strata ranging from semi-proletarians to landed proprietors. Consequently the peasants can have no guiding role in politics but can follow only one or the other of the two major contenders for power, i.e., the capitalists or the workers. Capitalists have traditionally used the peasants as a buffer against the workers and the Bolsheviks set out to change that situation.

Toward that end they developed the transitional slogan, "For a Workers and Peasants Government." Their propaganda around the slogan was not directed to the whole peasantry, however. It was presented in such a way as to differentiate the poor peasants from the rich ones and to draw the former toward an alliance with the workers.

Since the peasantry constitute a major section of the population in most countries, outweighing all other potential working class allies, the slogan was used generally on an international scale. But an exception was made where the peasantry represents a less substantial social force. In Britain, for example, with the agrarian sector amounting to only a minor factor, the slogan was truncated to read "For a Workers Government."

When the question was taken up by the SWP in 1938, it was considered in the light of conditions then existing in the United States. Rural families constituted around one-fourth of the total population. Their relative political weight—as compared with other potential allies of the workers—was correspondingly substantial. In these circumstances we adopted the slogan "For a Workers and Farmers Government," using the U.S. term for those who work the land.

Since then technological change and the growth of monopoly on the land have sharply altered the situation. The farm population, including all categories, has now dropped to about one-sixteenth of the total population and the trend continues downward. Farmers no longer constitute an especially significant force to be singled out above all other potential allies of the working class. A change in the slogan is therefore indicated.

It is neither necessary nor practical to list all potential allies of the workers in the transitional slogan. The key factor is the idea of a struggle for power led by the workers and supported by all their allies. These allies can be mentioned specifically and their political roles discussed in our propaganda put forward around the central concept of a workers government.

For these reasons it is recommended that the slogan be changed to "For a Workers Government."

EXPAND YOUR REVOLUTIONARY LIBRARY

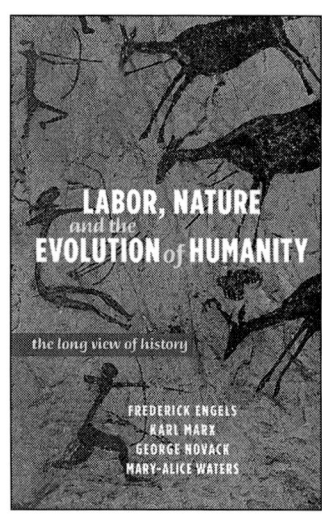

Labor, Nature, and the Evolution of Humanity
The Long View of History
FREDERICK ENGELS
KARL MARX
GEORGE NOVACK
MARY-ALICE WATERS

Without understanding that social labor, transforming nature, has driven humanity's evolution for millions of years, working people are unable to see beyond the capitalist epoch of class exploitation that warps all human relations, ideas, and values. Only the revolutionary conquest of state power by the working class can open the door to a world free of capitalist exploitation, degradation of nature, subjugation of women, racism, and war. A world built on human solidarity. A socialist world. $12. Also in Spanish and French.

Women in Cuba: The Making of a Revolution within the Revolution
VILMA ESPÍN
ASELA DE LOS SANTOS
YOLANDA FERRER

The integration of women in the ranks and leadership of the Cuban Revolution was intertwined with the proletarian course of the leadership of the revolution from the start. This is the story of that revolution and how it transformed the women and men who made it. $17. Also in Spanish, Farsi, and Greek.

Cosmetics, Fashions, and the Exploitation of Women
JOSEPH HANSEN, EVELYN REED MARY-ALICE WATERS

How big business reinforces women's second-class status and uses it to rake in profits. Where does women's oppression come from? How has the entry of millions of women into the workforce strengthened the battle for emancipation, still to be won? $12. Also in Spanish, Farsi, and Greek.

Lenin's Final Fight
Speeches and Writings, 1922–23
V.I. LENIN

In 1922 and 1923, V.I. Lenin, central leader of the world's first socialist revolution, waged what was to be his last political battle—one that was lost following his death. At stake was whether that revolution, and the international communist movement it led, would remain on the revolutionary proletarian course that brought workers and peasants to power in October 1917. $17. Also in Spanish, Farsi, and Greek.

Opening Guns of World War III: Washington's Assault on Iraq
JACK BARNES

The murderous assault on Iraq in 1990–91 heralded increasingly sharp conflicts among imperialist powers, growing instability of capitalism, and more wars. Also includes:

1945: When US Troops Said No! by Mary-Alice Waters

Lessons from the Iran-Iraq War by Samad Sharif

In *New International* no. 7. $14. Also in Spanish, French, and Farsi.

Socialism and Man in Cuba
ERNESTO CHE GUEVARA
FIDEL CASTRO

"Man truly reaches his full human condition when he produces without being compelled by physical necessity to sell himself as a commodity," wrote Guevara in 1965. $5. Also in Spanish, French, Farsi, and Greek.

Art and Revolution
Writings on Literature, Politics, and Culture
LEON TROTSKY

"Art can become a strong ally of revolution only insofar as it remains faithful to itself," wrote Trotsky in 1938. $15

The History of the Russian Revolution

LEON TROTSKY

How, under Lenin's leadership, the Bolshevik Party led millions of workers and farmers to overthrow the state power of the landlords and capitalists in 1917 and bring to power a government that advanced their class interests at home and worldwide. Unabridged, 3 vols. in one. Written by one of the central leaders of that socialist revolution. $30. Also in French and Russian.

Maurice Bishop Speaks
The Grenada Revolution and Its Overthrow, 1979–83

The triumph of the 1979 revolution in the Caribbean island of Grenada under the leadership of Maurice Bishop gave hope to millions throughout the Americas. Invaluable lessons from the workers and farmers government destroyed by a Stalinist-led counterrevolution in 1983. $20

Malcolm X: The Last Speeches

"Any kind of movement for freedom of Black people based solely within the confines of America is absolutely doomed to fail." Speeches and interviews from the last two years of his life. $12

The Low Point of Labor Resistance Is Behind Us
The Socialist Workers Party Looks Forward

JACK BARNES
MARY-ALICE WATERS
STEVE CLARK

The global order imposed by victors of the inter-imperialist slaughter of World War II is shattering, with explosive ramifications for workers and farmers worldwide. A long retreat by the working class and unions has come to an end. More and more workers of all ages, skin colors, and both sexes are saying, "Enough is enough!" This book highlights opportunities ahead for class-conscious workers to forge a labor party built on the unions. And a mass proletarian vanguard able to lead the struggle to end capitalist rule, opening a future for humanity. $10. Also in Spanish and French.

Capitalism and the Transformation of Africa
Reports from Equatorial Guinea

MARY-ALICE WATERS, MARTÍN KOPPEL

Describes how, as Equatorial Guinea is pulled into the world market, both a capitalist class and a working class are being born. Also documents the work of volunteer Cuban health-care workers there—an expression of the living example of Cuba's socialist revolution. $10. Also in Spanish and Farsi.

Pathfinder Press **accessible ebooks** for the blind, those with low vision, or other challenges reading print books

For a list of current accessible titles, go to: pathfinderpress.com/collections/books-for-the-blind.

Visit bookshare.org for information on how to sign up.

PATHFINDERPRESS.COM

RELATED READING

We Are Heirs of the World's Revolutions
Speeches from the Burkina Faso Revolution, 1983–87
THOMAS SANKARA

How peasants and workers in this West African country established a popular revolutionary government and began to fight hunger, illiteracy, and economic backwardness imposed by imperialist domination. They set an example not only for workers and small farmers in Africa, but their class brothers and sisters the world over. $10. Also in Spanish, French, and Farsi.

The Fight for a Workers and Farmers Government in the United States
JACK BARNES

The shared exploitation of workers and working farmers by banking, industrial, and commercial capital lays the basis for their alliance in a revolutionary fight for a government of the producers. In *New International* no. 4. $14

Their Trotsky and Ours
JACK BARNES

To lead the working class in a successful revolution, a mass proletarian party is needed whose cadres, well beforehand, have absorbed a world communist program, are proletarian in life and work, derive deep satisfaction from doing politics, and have forged a leadership with an acute sense of what to do next. This book is about building such a party. $12. Also in Spanish, French, and Farsi.

Renewal or Death: Cuba's Rectification Process
FIDEL CASTRO

Two speeches by Fidel Castro registering a turning point in the Cuban Revolution in the late 1980s known as rectification. It marked a turn away from copying many political and economic policies from the Soviet Union, and a revitalization of working-class methods, popular mobilization, and voluntary work. Introduction by Mary-Alice Waters. In *New International* no. 6. $14. Also in French.

Cuba's Internationalist Foreign Policy
Speeches, vol. 1, 1975–80
FIDEL CASTRO

Cuba's foreign policy, says Castro, starts "with the subordination of Cuban positions to the international needs of the struggle for socialism and national liberation." Speeches from 1975–80 on solidarity with Angola, Vietnam, the Nicaragua and Grenada revolutions, and more. $23

The Workers and Farmers Government
JOSEPH HANSEN

How experiences in post–World War II revolutions in Yugoslavia, China, Algeria, and Cuba enriched communists' theoretical and practical understanding of revolutionary governments of the workers and farmers. "What is involved is governmental power," writes Hansen, "the possibility of smashing the old structure and overturning capitalism." $5

Marxism and the Working Farmer
DOUG JENNESS, FIDEL CASTRO

Includes "American Agriculture and the Working Farmer" by Doug Jenness, "The Peasant Question in France and Germany" by Frederick Engels, "Theses on the Agrarian Question" by V.I. Lenin, and "Cuba's Agrarian Reform" by Fidel Castro. $6

CAPITALIST CRISIS AND THE FIGHT FOR WORKERS POWER

The Transitional Program for Socialist Revolution
LEON TROTSKY

The Socialist Workers Party program, drafted by Trotsky in 1938, still guides the SWP and communists the world over. The party "uncompromisingly gives battle to all political groupings tied to the apron strings of the bourgeoisie. Its task—the abolition of capitalism's domination. Its aim—socialism. Its method—the proletarian revolution." $17. Also in Farsi.

Are They Rich Because They're Smart?
Class, Privilege, and Learning under Capitalism
JACK BARNES

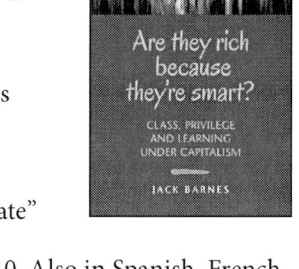

Exposes growing class inequalities in the US and the self-serving rationalizations of well-paid professionals who think their "brilliance" equips them to "regulate" working people, who don't know what's in our own best interest. $10. Also in Spanish, French, Farsi, and Arabic.

The Clintons' Anti-Working-Class Record
Why Washington Fears Working People
JACK BARNES

What working people need to know about the profit-driven course of Democrats and Republicans alike over the last three decades. And the political awakening of workers seeking to understand and resist the capitalist rulers' assaults. $10. Also in Spanish, French, Farsi, and Greek.

The Jewish Question
A Marxist Interpretation
ABRAM LEON

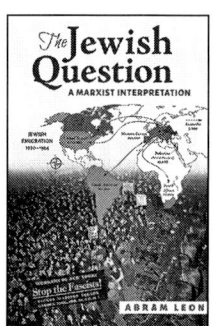

Why is Jew-hatred still raising its ugly head? What are its class roots—from antiquity through feudalism, to capitalism's rise and current crises? Why is there no solution under capitalism? The author, Abram Leon, was killed in the Nazi gas chambers. Revised translation, new introduction, and 40 pages of illustrations and maps. $17. Also in Spanish and French.

Cuba and the Coming American Revolution
JACK BARNES

This is a book about the struggles of working people in the imperialist heartland, the youth attracted to them, and the example set by the Cuban people that revolution is not only necessary—it can be made. It is about the class struggle in the US, where the revolutionary capacities of workers and farmers are today as utterly discounted by the ruling powers as were those of the Cuban toilers. And just as wrongly. $10. Also in Spanish, French, and Farsi.

Teamster Rebellion
FARRELL DOBBS

The 1934 strikes that won union recognition for truckers and warehouse workers in Minneapolis and helped pave the way for the working-class social movement that built the industrial unions. The first of four volumes by a central leader of these battles. $16. Also in Spanish, French, Farsi, and Greek.

The Communist Manifesto
KARL MARX AND FREDERICK ENGELS

Communism, say the founding leaders of the revolutionary workers movement, is not a set of ideas or preconceived "principles" but workers' line of march to power, springing from a "movement going on under our very eyes." $5. Also in Spanish, French, Farsi, and Arabic.

PATHFINDERPRESS.COM

DEFENDING WORKERS' RIGHTS

Socialism on Trial
TESTIMONY AT MINNEAPOLIS SEDITION TRIAL
James P. Cannon

The revolutionary program of the working class, presented in response to frame-up charges of "seditious conspiracy" in 1941, on the eve of US entry into World War II. The defendants were leaders of the Minneapolis labor movement and the Socialist Workers Party. $15. Also in Spanish, French, and Farsi.

Cointelpro
THE FBI'S SECRET WAR ON POLITICAL FREEDOM
Nelson Blackstock

An in-depth look at the 1960s and '70s covert FBI disruption and counterintelligence program—code-named COINTELPRO. Contains reproductions of FBI documents released through the Socialist Workers Party suit against government spying. $15

50 Years of Covert Operations in the US
WASHINGTON'S POLITICAL POLICE AND
THE AMERICAN WORKING CLASS
Larry Seigle, Farrell Dobbs, Steve Clark

How class-conscious workers have fought against the drive to build the "national security" state essential to maintaining capitalist rule. $10. Also in Spanish and Farsi.

FBI on Trial
THE VICTORY IN THE SOCIALIST WORKERS PARTY SUIT AGAINST GOVERNMENT SPYING
Margaret Jayko

The record of a historic victory in the fight for political rights, including the 1986 federal court ruling against government spying and excerpts from trial testimony by SWP leaders Farrell Dobbs and Jack Barnes. $17